276
Topics in Current Chemistry

Topics in Current Chemistry
Recently Published and Forthcoming Volumes

Creative Chemical Sensor Systems
Volume Editor: Schrader, T.
Vol. 277, 2007

In situ NMR Methods in Catalysis
Volume Editors: Bargon, J., Kuhn, L. T.
Vol. 276, 2007

Sulfur-Mediated Rearrangements II
Volume Editor: Schaumann, E.
Vol. 275, 2007

Sulfur-Mediated Rearrangements I
Volume Editor: Schaumann, E.
Vol. 274, 2007

Bioactive Conformation II
Volume Editor: Peters, T.
Vol. 273, 2007

Bioactive Conformation I
Volume Editor: Peters, T.
Vol. 272, 2007

Biomineralization II
Mineralization Using Synthetic Polymers and Templates
Volume Editor: Naka, K.
Vol. 271, 2007

Biomineralization I
Crystallization and Self-Organization Process
Volume Editor: Naka, K.
Vol. 270, 2007

Novel Optical Resolution Technologies
Volume Editors:
Sakai, K., Hirayama, N., Tamura, R.
Vol. 269, 2007

Atomistic Approaches in Modern Biology
From Quantum Chemistry
to Molecular Simulations
Volume Editor: Reiher, M.
Vol. 268, 2006

Glycopeptides and Glycoproteins
Synthesis, Structure, and Application
Volume Editor: Wittmann, V.
Vol. 267, 2006

Microwave Methods in Organic Synthesis
Volume Editors: Larhed, M., Olofsson, K.
Vol. 266, 2006

Supramolecular Chirality
Volume Editors: Crego-Calama, M.,
Reinhoudt, D. N.
Vol. 265, 2006

Radicals in Synthesis II
Complex Molecules
Volume Editor: Gansäuer, A.
Vol. 264, 2006

Radicals in Synthesis I
Methods and Mechanisms
Volume Editor: Gansäuer, A.
Vol. 263, 2006

Molecular Machines
Volume Editor: Kelly, T. R.
Vol. 262, 2006

Immobilisation of DNA on Chips II
Volume Editor: Wittmann, C.
Vol. 261, 2005

Immobilisation of DNA on Chips I
Volume Editor: Wittmann, C.
Vol. 260, 2005

In situ NMR Methods in Catalysis

Volume Editors: Joachim Bargon · Lars T. Kuhn

With contributions by

J. Bargon · R. Giernoth · L. Greiner · L. T. Kuhn
S. Laue · A. Liese · H. G. Niessen · K. Woelk · J. Wöltinger

Springer

The series *Topics in Current Chemistry* presents critical reviews of the present and future trends in modern chemical research. The scope of coverage includes all areas of chemical science including the interfaces with related disciplines such as biology, medicine and materials science. The goal of each thematic volume is to give the nonspecialist reader, whether at the university or in industry, a comprehensive overview of an area where new insights are emerging that are of interest to a larger scientific audience.
As a rule, contributions are specially commissioned. The editors and publishers will, however, always be pleased to receive suggestions and supplementary information. Papers are accepted for *Topics in Current Chemistry* in English.
In references *Topics in Current Chemistry* is abbreviated Top Curr Chem and is cited as a journal.
Visit the TCC content at springerlink.com

ISSN 0340-1022

DOI 10.1007/978-3-540-71427-9

ISBN 978-3-642-09060-8 e-ISBN 978-3-540-71427-9

Springer is a part of Springer Science+Business Media

springer.com

Softcover reprint of the hardcover 1st edition 2007

Cover design: WMXDesign GmbH, Heidelberg

Printed on acid-free paper 02/3180 YL – 5 4 3 2 1 0

Volume Editors

Prof. Dr. Joachim Bargon

Institute of Physical
& Theoretical Chemistry
University of Bonn
Wegelerstrasse 12
53115 Bonn, Germany
bargon@uni-bonn.de

Lars T. Kuhn

Physical & Theoretical Chemistry
Laboratory
South Parks Road
Oxford OX1 3QZ
United Kingdom
lars.kuhn@chem.ox.ac.uk

Topics in Current Chemistry
Also Available Electronically

For all customers who have a standing order to Topics in Current Chemistry, we offer the electronic version via SpringerLink free of charge. Please contact your librarian who can receive a password or free access to the full articles by registering at:

springerlink.com

If you do not have a subscription, you can still view the tables of contents of the volumes and the abstract of each article by going to the SpringerLink Homepage, clicking on "Browse by Online Libraries", then "Chemical Sciences", and finally choose Topics in Current Chemistry.

You will find information about the

- Editorial Board
- Aims and Scope
- Instructions for Authors
- Sample Contribution

at springer.com using the search function.

Preface

A chemist, faced with the problem of determining the mechanism of a chemical reaction, tries to identify a set of reactions that will account for the observed behavior: Ideally, a small set of known reactions should describe in great detail exactly what takes place at each stage of a chemical transformation. The fact that many reactions proceed in a stepwise fashion can most convincingly be demonstrated if intermediate species can be isolated and shown to proceed to the same products under otherwise identical reaction conditions. An intermediate is the reaction product of each of these steps, except for the last one that forms the final **product.** Some intermediates are stable compounds in their own right; some others, however, are so reactive that their isolation is not possible.

Occasionally, evidence for the existence of short-lived intermediates may be obtained, in particular by spectroscopic observation. The latter may allow a direct observation or an indirect inference from unusual phenomena occurring in the reaction products during in situ investigations of their corresponding chemical reactions. In NMR spectroscopy, for example, transient emission and enhanced absorption lines may be observed, and one is inclined to believe that there is a universal and unambiguous reason for their appearance. This is not necessarily the case, however, since this seemingly identical phenomenon may have a strikingly different origin: During free radical reactions, a phenomenon called chemically induced dynamic nuclear polarization (CIDNP) may give rise to virtually the same effect as occasionally observed during homogeneous (and possibly even heterogeneous) hydrogenations: The latter phenomenon, called parahydrogen-induced polarization (PHIP), has a completely different physical basis. It was first noticed twenty years later than CIDNP and occurs if there is an imbalance of the two spin isomers of symmetric molecules such as dihydrogen when hydrogenating unsaturated compounds using appropriate catalysts. These two effects, if not differentiated properly, can cause misinterpretations of reaction mechanisms, as occurred initially when their different origins had not yet been understood appropriately.

In this volume, both phenomena, CIDNP and PHIP, will be described and typical applications outlined. Apart from providing interesting insights into catalytic and free radical reaction mechanisms, these examples of chemical reaction-assisted nuclear polarization together with their associated signal and sensitivity enhancement of nuclear magnetic resonance are not only of interest in chemistry, but are also rapidly gaining significance in medicine and biochemistry as a potential means to boost the sensitivity of magnetic resonance imaging (MRI). “Hyperpolarizing” ^{13}C-nuclei for example, has been demonstrated to provide access to fast angiography. Likewise, following the distribution and metabolism of ^{13}C-hyperpolarized compounds might provide an alternative approach to radioactively labeled targets, augmenting or replacing imaging methods of nuclear medicine, such as positron emission tomography (PET). When superimposed with ^{1}H-MRI data, for example, the corresponding resolution of this approach may exceed that of conventional PET studies, even though at present the sensitivity of PET still substantially outshines that of MRI.

Additional chapters dealing with special boundary conditions such as ionic liquids, supercritical solvents, and biocatalysis are also highlighted and augment this account of in situ NMR methods in catalysis. The authors of these chapters were associated with the University of Bonn in one form or another early on in their careers.

University of Bonn
March 2007

Joachim Bargon
Lars T. Kuhn

Contents

Contents of Volume 246

New Techniques in Solid-State NMR

Volume Editor: Jacek Klinowski
ISBN: 978-3-540-22168-5

Top Curr Chem (2007) 276: 1–23
DOI 10.1007/128_016

Published online: 18 November 2005

Homogeneous Catalysis in Ionic Liquids

Ralf Giernoth

Institute of Organic Chemistry, University of Cologne, Greinstr. 4, 50939 Cologne, Germany
ralf.giernoth@uni-koeln.de

Abstract In the past 15 years, ionic liquids have become an alternative reaction medium for organic transformations, especially for transition metal catalysis. Their unique properties make them ideal solvents for "green" industrial processes: they are polar, thus exhibiting high solubility for a large variety of substrates and catalysts, they are immiscible with many organic solvents, and they do not evaporate in high vacuum. Many of their physicochemical properties are changed substantially by variation of the cation and the anion; thus, they are "tunable" to the desired reaction. This review focuses on the general concepts that are applicable to ionic liquids as reaction media. This knowledge is intended to enable the reader to use ionic liquids advantageously for their chemistry. In the second part, some recent examples of successful ionic liquid solvent chemistry are pointed out and discussed in more detail.

Keywords Biphasic catalysis · Green chemistry · Ionic liquids · Transition metal catalysis

Abbreviations

bdmim	1-Butyl-2,3-dimethylimidazolium
bmim	1-Butyl-3-methylimidazolium
bmpy	1-Butyl-4-methylpyridinium
BTA	Bis(trifluromethylsulfonyl)amide
C_{10}mim	1-Decyl-3-methylimidazolium
emim	1-Ethyl-3-methylimidazolium
IL	Ionic liquid
RTIL	Room-temperature ionic liquid
$scCO_2$	Supercritical carbon dioxide

1 Introduction

Since the beginning of the 1990s, chemistry has rapidly started to evolve towards more and more environmentally benign processes for synthetic applications. Nowadays, "green chemistry" [1] is an important keyword in chemical research. Analysis of the factors that have the strongest influence on environmental issues in a chemical process frequently comes down to the influence of the solvents: heating and cooling of the reaction mixture consume energy, solubility problems reduce the choice of solvents and the efficiency of processes, the toxicity of solvent vapours are of concern, recycling of toxic or expensive catalysts is often difficult, etc. Therefore, many attempts have been made to substitute classical organic solvents with novel reaction media, tailor-made for the specific task they are needed for. Among the more frequent examples are perfluorinated solvents [2], supercritical carbon dioxide ($scCO_2$) [3], and, of course, ionic liquids (ILs).

The field of ILs as reaction media in organic synthesis has evolved rapidly in the last 15 years—Fig. 1 shows the number of publications with the keyword "ionic liquids" as a function of the publication year. Consequently, there are numerous reviews in the literature already, giving a comprehensive overview about the topic as well as focusing on certain specialities [4–11]. In 2003, the first comprehensive book, *Ionic liquids in synthesis* [12], was published, which

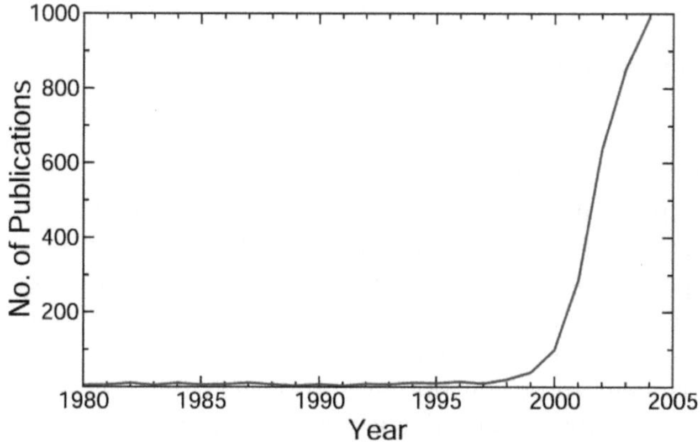

Fig. 1 Numbers of publications with ionic liquids as the topic (Source: SciFinder Scolar)

contains the state of the art when it comes down to ILs. Therefore, there is no urgent need for another comprehensive review at the present date. This article will, although giving a general overview about the topic, focus on recent developments of the last 5 years and on certain specific issues in transition metal catalysis that the author perceives as especially important and interesting to the public. The goal is to give the reader a "feeling" for what is possible and what we can expect from IL chemistry in the near future.

2
A Short History of Ionic Liquids

The "modern history" of ILs as reaction media for homogeneous transition metal catalysis really started in 1990, when Chauvin et al. [13] developed weakly acidic chloroaluminate melts for dimerisation and polymerisation reactions. Shortly afterwards, in 1992, Wilkes and Zaworotko [14] developed a totally new class of IL solvents with tetrafluoroborate and hexafluorophosphate anions. These melts were stable towards hydrolysis and showed a high tolerance towards molecules with reactive functional groups. From that time on, room-temperature ILs (RTILs) were available as general reaction media for synthetic chemistry.

Of course, organic salts that were liquid at room temperature were no invention of the 1990s. In fact, the very first IL (although this term was introduced much later) was described in 1914 [15]: ethylammonium nitrate $[EtNH_3]NO_3$, exhibiting a melting point of just 12 °C. In the years starting from 1948, RTILs were mainly used as a medium for electrochemical applications, until in the mid-1980s Seddon and Hussey started using chloroaluminate melts as media for organic synthesis [16].

3
Classification of Ionic Liquids

What is an IL? If we want to classify these novel reaction media, we first have to define what is to be called an IL.

An IL is a (partially) organic salt that is liquid at room temperature or reaction temperature (RTILs versus ILs). As Welton [9] has pointed out, there is nothing special about room temperature apart from the fact that this is the temperature that rooms happen to be at, so this definition is only of limited use—especially since many reactions are run at elevated temperatures. Wasserscheid and Keim [8] have proposed calling an organic salt "IL" if it is liquid below 100 °C, and this is indeed now one of the most widely accepted definitions. Where liquidity at room temperature is the key feature, the term RTIL has found widespread use in the chemical community.

In contrast to what we would call "molten salts", a term that is always associated with high melting points and a quite corrosive medium, ILs are low-melting, quite unreactive, non-corrosive, and therefore suitable as reaction media (i.e., solvents).

The classification of ILs can, of course, be based on any of their physicochemical properties, such as melting point, liquidus range, viscosity, acidity/basidity, and density. These properties will be the topic of the next section. In practical terms, the classification of ILs is normally based on their chemical structure, i.e., on the composition of their cations in combination with their anions.

Most ILs consist of a cation that bears a quarternary ammonium or phosphonium centre (Scheme 1), although systems based on arsonium, antimonium, sulfonium, etc. have also been described. This quarternary centre is normally substituted in an unsymmetric fashion, i.e., at least one of the substituents differs from all the others. This reduction of symmetry is necessary to prevent the salt from crystallising easily, widening the liquidus

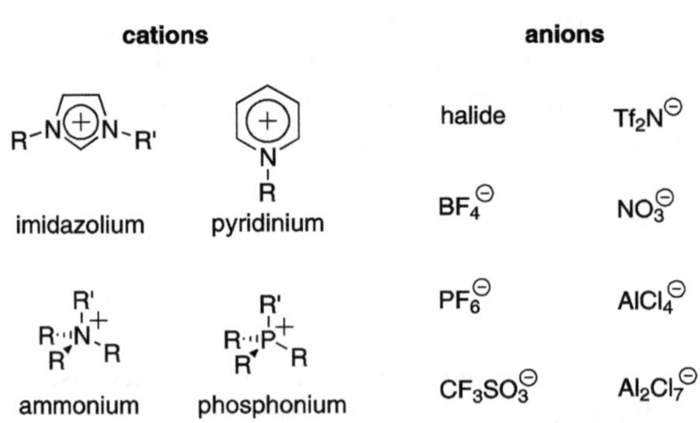

Scheme 1 Commonly used cations and anions that combine to form ionic liquids

range. In the case of simple ammonium and phosphonium salts, for example, this can be achieved by having three identical and one different substituent on the quarternary centre. Nowadays, even more common are cations based on unsymmetrically substituted imidazolium, pyridinium or pyrrolidinium moieties.

Virtually every imaginable anion can be used as the counterion—every one of these showing various advantages and disadvantages. In the beginning of IL solvent chemistry, mixed anions of the chloroaluminate type, i.e., mixtures of Cl^- and $AlCl_3$, were very common, but since their properties change with composition [8, 9] and since they are not stable towards hydrolysis, these are only of limited use for transition metal catalysis and therefore beyond the scope of this review. Among the most common anions that are covered here are the halides, triflate, PF_6^-, BF_4^-, and Tf_2N^- (sometimes also called bistriflamide or bis(trifluoromethylsulfonyl)imide (BTA).[1]

Since the names of commonly used ILs can be quite long and very unhandy, a shortened nomenclature has become widely accepted in the IL community. Thus, 1-ethyl-3-methylimidazolium bromide becomes [emim]Br, 1-decyl-3-methylimidazolium tetrafluoroborate becomes $[C_{10}mim]BF_4$, 1-butyl-4-methylpyridinium bis-(trifluromethylsulfonyl)amide becomes [bmpy] Tf_2N or [bmpy]BTA and so on. If unsure, please refer to the list of abbreviations at the beginning of this review.

4 Properties of Ionic Liquids

4.1 General Properties

The most prominent property of all ILs is the fact that they do not have any measurable vapour pressure—obviously so, since they are salts and therefore are completely composed of cations and anions. (Recently, various reports of distillable ILs have appeared in the literature, e.g., [17]. At the present state, I leave it to the reader to make up his or her own mind about this topic.) This fact gives rise to one of the major advantages of using ILs in synthesis and their label of being "green": no vapour pressure means no volatile solvent and no toxic solvent vapours. Additionally, the product of a reaction run in an IL can be obtained by simply distilling it off the solvent while the catalyst stays "immobilised" in the ionic phase—the same holds

[1] There is an ongoing disagreement in the chemical literature about the question whether the Tf_2N^- anion is to be called an imide or an amide. From an inorganic point of view (and this anion is surely inorganic), salts of the general formula M^+NR_2 are amides, $M_2{}^+NR$ salts are imides and $M_3{}^+N$ salts are nitrides. It seems therefore quite obvious to the author that Tf_2N^- has got to be bis(trifluoromethylsulfonyl)amide

for extraction processes. And, important for transition metal catalysis, which is often air- and moisture-sensitive, the solvent can be heated in high vacuum prior to reaction, making it easy to provide an inert solvent with little effort.

The second important property common to all ILs is their potential to dissolve a wide variety of organic and inorganic compounds. Although they are not, as one could suspect, "superpolar" (their polarity is typically in the range of acetonitrile and methanol [18, 19]), the ionic structure enables them to dissolve polar substances (since they are ions) as well as unpolar ones (e.g., via non-polar interactions through an aliphatic side chain). Therefore, it is possible to dissolve a quite diverse range of reactants in a single phase, enabling reactivities that are impossible in conventional organic solvents.

4.2 Variable Properties

Apart from the few physicochemical properties that are common to all ILs, the majority of them vary substantially with the choice of cation and anion. ILs are thus frequently called "designer solvents", for it is possible to choose one that exhibits the exact properties needed for a given reaction. In fact, this variability is one of the major advantages of IL chemistry: there is no such thing as *the* IL but there are potentially millions! Therefore, often-heard questions like "does my reaction work in ILs?" should actually read "which IL could be suitable for my reaction?" For transition metal catalysis this means that, in addition to the catalyst, the ligands, and the substrates, the solvent can be optimised in a very defined fashion. Selected examples for this kind of solvent tuning are given later in this article; some of the tunable properties are discussed in the following.

4.2.1 Melting Points and Liquidus Range

By definition (Sect. 3), the melting point of an IL lies below 100 °C (below room temperature for an RTIL, respectively). With a given cation, the choice of anion has a strong effect on the melting point [20] (Table 1). Coordinating and hydrophilic anions like the halides lead to high melting points, whereas weakly coordinating and hydrophobic anions result in low melting points. By variation of the alkyl chain length in the cation, fine-tuning of the melting point can be achieved [8, 20]. Symmetrically substituted cations can crystallise easily and therefore often lead to "ionic solids" (i.e., high melting points). Low symmetry in substitution can prevent easy crystallisation, resulting in low melting points. The longer the alkyl chain, the lower is the melting point, but only up to a certain extent (rule of thumb for imidazolium cations: C_8 gives the lowest melting points [20]). Beyond that, prolongation

Table 1 Melting points of selected ionic liquids [16]

Ionic liquid	Melting point (°C)
[emim]Cl	87 °C
[emim]PF_6	62
[emim]NO_3	38
[emim]BF_4	15
[emim]Tf_2N	– 3

of the alkyl chain raises the melting point again. Additionally, a good distribution of the positive charge over a number of atoms seems to favour low melting points [8].

The liquidus range is limited at the lower end, of course, by the melting point of the IL. Below that, the IL either crystallises, often firstly as a liquid crystal, or transforms to a glass state. The upper limit is obviously not given by the boiling point (for the IL exhibits no vapour pressure) but by the thermal decomposition temperature. For the imidazolium-based ILs, the thermal decomposition range is between 250 and 500 °C. Therefore, by proper choice of cation and anion, one can easily obtain a liquidus range of 350 °C.

4.2.2 Viscosity

Varying strongly with a change of cation and anion, the viscosity of ILs can range from "very viscous" (almost like honey) to "not too viscous" (like, e.g., water) [21]. The choice of anion has the strongest effect here, but tiny amounts of water (even in the parts-per-million range) that are almost always present in the IL also exhibit a very strong effect on lowering the viscosity. For numerous data, see Ref. [21]. Directly connected with this is the self-diffusion coefficient, which in itself influences the dynamics and kinetics of transition metal catalysed reactions in these media.

4.2.3 Polarity

As mentioned earlier, one could expect that ILs exhibit some kind of superpolarity, but this is not the case. Their polarity is normally in the range of acetonitrile or methanol, depending on the method used for its determination [18, 19, 22]. Still, there is an ongoing discussion about the "right" way to determine solvent polarities, especially for ILs [8]. It has been shown, for example, that with use of ILs as the stationary phase in gas chromatography, strong interactions with polar substrates as well as with non-polar substrates could be detected [23]. This ambiguity also reflects in their ability to dissolve polar as well as non-polar molecules.

4.2.4 Phase Behaviour

Probably the most important property from a synthetic chemist's point of view is the phase behaviour of ILs. ILs can be hydrophilic, completely hydrophobic, or partially miscible with water, giving either one or two phases. The same is true for their miscibility with all other molecular ("organic") solvents. This feature can be used as an advantage: it is possible to work in mixed IL/water/organic solvent systems as well as to do biphasic catalysis or extraction of the desired product from the IL phase with an organic solvent.

A very nice example of using the phase behaviour of an IL as a process advantage was demonstrated by Mathews et al. [24] (Scheme 2): when running the Suzuki–Miyaura coupling reaction of bromobenzene and phenylboronic acid in [bmim]BF_4, the palladium catalyst and the reaction partners are completely dissolved in the IL. An aqueous solution of sodium carbonate which is miscible with the ionic phase is added. The reaction is started by increasing the temperature to 110 °C. The by-products of the reaction are preferentially soluble in water, changing the miscibility of the water phase with the IL. Therefore, the initially homogeneous solution is spontaneously split into two phases. The reaction product can then be extracted from the IL using diethyl ether. The mixture ends up as a triphasic system: the ether phase contains the product, the water phase (which can easily be separated from the rest) contains all the salt by-products, and the ionic phase contains only the still-active

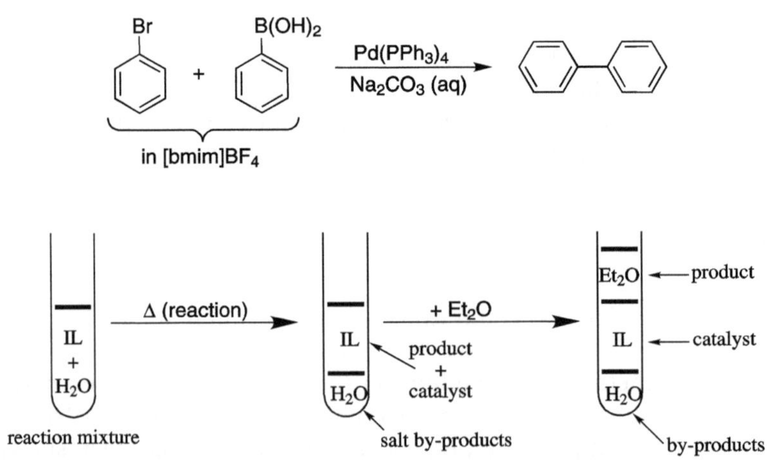

Scheme 2 Phase behaviour of the ionic liquid (*IL*)/water system running a Suzuki–Miyaura reaction [24]

catalyst. This catalyst can easily be recycled without tedious purification and can be used for subsequent reactions.

4.2.5 Miscellaneous Properties

Naturally, many other physicochemical properties vary (and can "actively" be varied) by the choice of anion and cation. Among those are reactivity-critical properties, like gas solubilities (e.g., for hydrogenations, hydroformylations, and oxidations), density, etc. For an extensive overview, see Ref. [12].

5 Synthesis of Ionic Liquids

The synthesis of ILs normally consists of two major steps. In the first step, the desired cation has to be generated, usually by direct alkylation/quaternisation of a nitrogen or phosphorus atom. In the second step, the anion resulting from the alkylation reaction can be exchanged for a different one [25]. Since imidazolium-based ILs have reached some kind of "standard" status in the IL community, I will focus on their synthesis in more detail (Scheme 3).

The cation is easily synthesised from readily available *N*-methylimidazole and the desired alkyl bromide or chloride by direct alkylation, often without the use of any solvent. The reaction with alkyl bromides is fast and very exothermic; thus, the *N*-methylimidazole has to be added dropwise to the alkyl bromide at 80 °C. On the downside, because of the exothermicity, the product is normally impure as a result of thermal degradation of the starting materials (especially *N*-methylimidazole). Alkylation with the alkyl chloride often leads to a cleaner product, but it takes much longer (often 3–4 days) until completion. In both cases the crude product can be purified by exposing it to high vaccuum at elevated temperature and subsequent recrystallisation (e.g., from acetonitrile/toluene).

Many of these halide salts are liquid below 100 °C. If a different anion is desired that is not accessible by direct alkylation, the halide needs to be ex-

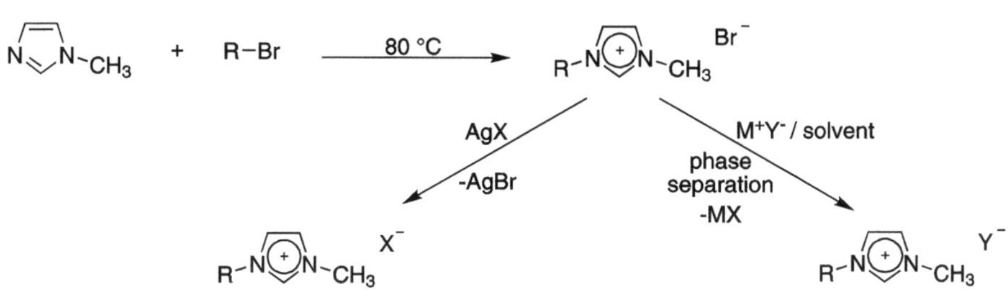

Scheme 3 Synthesis of imidazolium-based ionic liquids

changed in a second reaction step. This can be done by direct anion–anion exchange from the desired acid or salt. If, for example, the BF_4^- anion is needed, the exchange is readily done with $NaBF_4$ in acetone or HBF_4 in water and subsequent extraction of the IL with dichloromethane. Most of the ILs are colourless once they are pure (Sect. 6). If they are obtained strongly coloured (normally yellow to orange), they can be purified further, for example with activated charcoal and subsequent column chromatography [26]. Our own detailed synthetic procedure can be found in Ref. [27].

6 Purity of Ionic Liquids

Purity is a major issue one has to deal with when using ILs as reaction media, especially for transition metal catalysis. Once the ILs have formed in the course of the synthesis, purification can become a nuisance, to say the least. Obviously, they cannot be distilled (although a few reports on "distillable ionic media" have appeared in the literature [17]), most of them cannot be recrystallised, and column chromatography via silica is tricky (like with most polar solvents). Chromatography via reversed silica is an option, albeit an extremely expensive one. Therefore, it is worthwhile to consider the potential impurities that will be part of the final product.

For transition metal catalysis, one major problem is halide impurities stemming from incomplete anion exchange. For imidazolium-based systems with medium side chains (C_4– C_8), these can be removed by extraction with water [28]. [emim]-based ILs are too soluble in water for this method of purification, while the ones with longer alkyl chains are amphiphilic, resulting in quite a robust foam when extracted with water and thus in difficulties with phase separation. For these cases, it may sometimes be necessary to remove any residual halides by titration with $AgBF_4$, which can also be, of course, quite expensive and may lead to silver impurities in the IL.

Gallo et al. [29] have systematically studied the influence of halide impurities on catalytic Michael addition reactions. They have found that the system is strongly sensitive to the amount of halides present in the IL, inhibiting the activity of the transition metal catalyst. Furthermore, the total amount of halide impurities in different IL batches is variable even if the same synthetic protocol is followed. For a palladium-catalysed copolymerisation of styrene and carbon monoxide, Klingshirn et al. [30] came to the same conclusions. Daguenet and Dyson [31] explained this fact as a consequence of the extremely weak interactions between the halide anion and the imidazolium cation, through which the dissociation of the halide from a transition metal complex can become thermodynamically disfavoured in ILs. In contrast, we ourselves were able to demonstrate that the efficiency of Suzuki-type reactions in BF_4^--based ILs is strongly *promoted* by the presence of fluoride anions (Giernoth, unpublished results) that present a common degradation product of these ILs when they come into contact with water [32, 33].

The second major purity problem is "colour". Most ILs are colourless in pure form, but in reality, they are more likely to be pale yellow to dark orange. The origin of this is still somewhat unclear, since these (often trace) impurities are not detectable via NMR or IR spectroscopy. Most likely, the colour is due to degradation of the starting material *N*-methylimidazole. Therefore, with a few precautions, colourless ILs can be obtained by (1) using freshly distilled *N*-methylimidazole for the synthesis, (2) performing the alkylation step at the most modest temperatures possible (i.e., avoiding overheating) under a protective atmosphere, and (3) by cleaning the final IL product through stirring with activated charcoal [26].

The third issue concerning purity is the amount of water present in the ILs. This is not only a problem for running reactions with water-sensitive compounds, but the amount of water can change the physical properties of an IL dramatically [28, 34]. Therefore, it is always advisable to dry ILs at elevated temperature in high vacuum with vigorous stirring overnight before using them. Stirring is crucial here because of high viscosities and because the water desorption takes place only via the surface of the liquid phase. In critical cases, the amount of water present can additionally be checked by IR spectroscopy [28] or, of course, by standard Karl Fischer titration.

7 Availability of Ionic Liquids

Since the first commercial supplier, Solvent Innovation [35], started doing business in 1999, a large variety of ILs have become commercially available at reasonable prices. Today, all of the major suppliers of fine chemicals have entered the market. Thus, it is easy for the fist-time or infrequent user of ILs to get hold of them. But still, ILs can be quite expensive, especially for being "just a solvent", depending on the anion and the purity needed. Therefore, frequent users will probably want to start synthesising larger quantities on their own, particularly in cases where very high grade purity is needed.

8 Concepts: Ionic Liquids in Homogeneous Catalysis

The major reason for using "novel" reaction media in homogeneous catalysis (and this holds for ILs as well as for supercritical fluids, perfluorinated solvents, etc.) is the desire to combine the advantages of heterogeneous catalysis (such as ease of catalyst separation and recycling) with the advantages of homogeneous catalysis (like high activities and selectivities, moderate reaction conditions, and the ability for ligand tuning) [36]. This section sums up the basic concepts and their advantages that have already been realised in using ILs as reaction media.

8.1 Biphasic Catalysis

With the wide choice of ILs available, it is often possible to find a biphasic IL/organic system for which the catalyst dissolves exclusively in the ionic phase while the reactants and/or the reaction products form a second phase or dissolve predominantly in a second organic reaction phase [5], or can exclusively be extracted with an organic solvent without any leaching of the catalyst. In this way, the transition metal catalyst is heterogenised in the ionic phase and can easily be separated from the reaction mixture while all the advantages of a homogeneous process remain.

Extraction with organic solvents from biphasic IL/water reactions can even lead to a triphasic system at the end of the reaction. When running Heck-type reactions in [bmim]PF_6, the standard palladium catalysts like $Pd(OAc)_2$ or $PdCl_2$ dissolve exclusively in this IL [37]. The reaction products can be extracted with cyclohexane, the salt by-products with water. Doing so simultaneously leads to a triphasic system (Scheme 4). The reaction product is in the organic phase, the by-products in the water phase (which can be disposed off), and the active catalyst in the ionic phase (which can easily be recycled). Recycling six times without loss of activity has been reported. Mathews et al. [24] reported similar phase behaviour for Suzuki–Miyaura reactions in [bmim]BF_4 with the additional feature that water is not used for extraction purposes only, but the whole reaction is run in a water/IL mixture (because the base needed for the reaction is not soluble in the IL). The process has already been described (Sect. 4.2.4, Scheme 2).

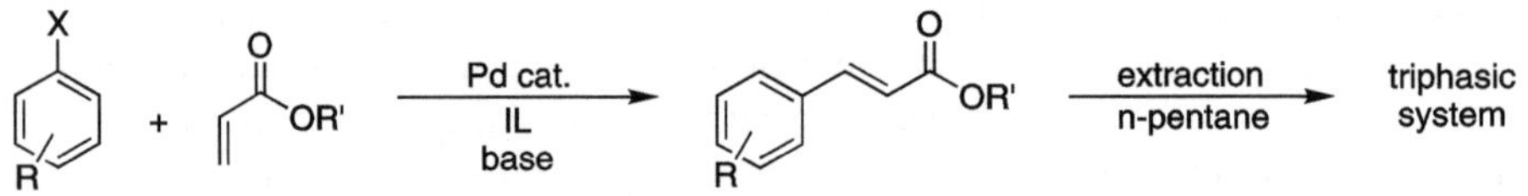

Scheme 4 Triphasic catalyst system for Heck reactions [37]

8.2 Monophasic Catalysis in "Inert" Ionic Liquids, or: Is the Ionic Liquid Just an "Innocent" Solvent?

There are numerous examples of effective transition metal catalysis in ILs—some selected examples will be discussed later in this review. The promoting effect of ILs can theoretically be due to at least two reasons.

First, the IL can act simply as an inert, weakly coordinating reaction medium. In such a medium, the transition metal catalyst would lack a tight solvent shell. A more-or-less "naked" catalyst would be able to show significantly higher activity compared with one that is dissolved in a coordinating solvent. In this case, the IL would simply be responsible for providing a medium with good solubility.

Second, the IL may as well take an active role in catalysis. The anions and the cations of the medium can participate in the catalytic cycle, influencing reactivities and selectivities.

The ongoing discussion about the "innocence" of ILs as solvents and the existence of a "general IL effect" is the focus of quite a few recent studies. Of course, the 2-proton of the imidazolium cation is acidic [38]. Xu et al. [39] have studied palladium-catalysed Heck reactions in [bmim]Br and [bmim]BF_4. In the bromide case, they were able to isolate imidazolylidene complexes of palladium and to demonstrate that these were active as catalyst precursors. Therefore, imidazolium-based ILs are prone to form imidazolylidene carbene complexes, especially in the presence of a transition metal or base (Scheme 5). The obvious question is whether these might be responsible for the often high activity of transition metal catalysts in these media.

Mathews et al. [40, 41] were able to demonstrate for their Suzuki–Miyaura-type reactions (Sect. 8.1) that a similar imidazolylidene palladium complex can form under the reaction conditions. However, they failed to prove that this complex is responsible for the high catalytic activity. By examining similar IL systems that are lacking any obvious carbogenic centre (e.g., pyridinium salts), we ourselves could show that imidazolylidene palladium complexes are (at least) not needed for high catalyst activity (Giernoth, unpublished results).

On the other hand, nickel(II) imidazolylidene complexes have been prepared as catalysts for olefin dimerisation reactions by McGuinness et al. [42]. They have been tested in toluene as well as in an imidazolium-based IL as the solvents. The catalysts were inactive in toluene, but highly active in the IL, which was interpreted in terms of catalyst stabilisation by the imidazolium cation of the solvent.

Finally, Aggarval et al. [43] observed an "unexpected side reaction" when running a base-catalysed Baylis–Hillman reaction in imidazolium-based ILs. The imidazolylidene (which is formed by base catalysis) reacts with the alde-

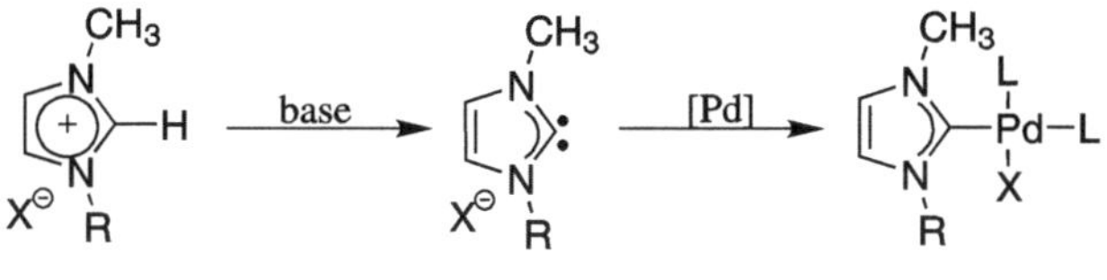

Scheme 5 Formation of Arduengo-type imidazolylidene carbene species from imidazolium salts

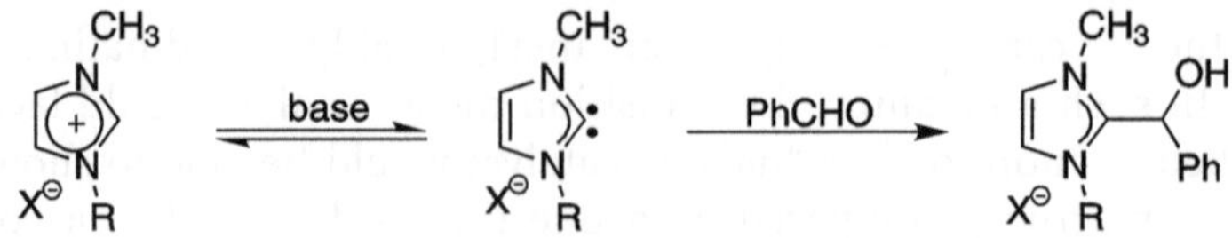

Scheme 6 Side reaction in the Baylis–Hillman reaction when run in imidazolium-based ILs [43]

hyde to form an adduct (Scheme 6). But, knowing about the acidity of the imidazolium 2-proton, we are hardly surprised by a side reaction of this type. All these, and probably many more, examples should be kept in mind when thinking of ILs as unreactive, "innocent" solvents.

8.3 Task-Specific ILs

Successful attempts to incorporate the active catalyst directly into the IL have recently been published. The groups of Dyson and Welton [44] have developed the IL [bmim]Co(CO)$_4$. By incorporation of a cobalt carbonyl complex as the anion, this IL shows catalytic activity in itself, for example in the debromination of 2-bromoketones (Scheme 7). The authors conclude that this general concept will be applicable to other catalytically active anions; analogous compounds with iron or manganese carbonyls are also liquid at room temperature. The obvious advantage of this approach is that there is no need for catalyst preparation, and separation of the catalyst from the reaction product is thus very easy. On the downside, it might turn out that transition-metal-containing ILs might be too expensive to be used as solvents for industrial applications, especially because of the high quantity of metal needed.

At ExxonMobil Research, Mehnert et al. [45, 46] have covalently bound imidazolium cations of ILs to the solid phase. This concept, which is referred to as "supported IL catalysis", has been demonstrated for a variety of catalytic reactions, including hydroformylations [45] and hydrogenations [47].

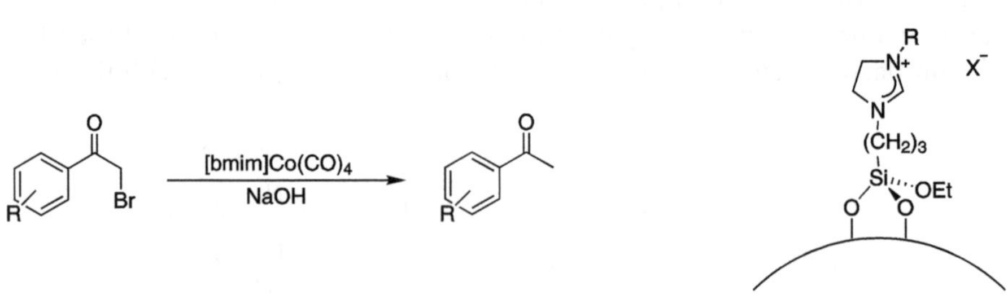

Scheme 7 Task-specific ionic liquids. *Left*: A catalytically active organometallic IL [44]. *Right*: Supported IL catalysis [45–47]

Riisager et al. [48] have followed a similar approach not by attaching ILs covalently bound to the solid phase, but by dissolving a catalyst in a thin IL film that is held on a porous solid by physisorption. This widely applicable technique has been termed supported IL-phase (SILP) catalysis by the authors. In this fashion, the advantages of ILs (the "IL effects") can be exploited using significantly reduced amounts of the expensive IL. Especially important for industrial purposes is the fact that the preferred traditional fixed-bed technology can be used. This approach makes the removal and recycling of the IL straightforward.

8.4
Process Engineering

Countless approaches have been followed to use ILs for "green" process engineering—for example, by immobilising transition metal catalysts in the ionic phase and recycling the catalytically active IL for subsequent reactions. One of the most successful techniques, aiming towards a "clean" and continuous process, is the combination of an IL phase with $scCO_2$. The general idea is to combine the advantages of $scCO_2$ as a reaction medium, like its high solvation potential for non-polar substrates, its non-toxicity, and its high self-diffusion coefficient [3], with the advantages of the IL medium. Many ILs show complete miscibility with $scCO_2$, and CO_2, once below the critical point, is just a non-toxic gas and thus easily removed from the reaction mixture. Additionally, CO_2 is highly soluble in most common ILs.

Brown et al. [49] have exploited the combination of IL and $scCO_2$ for ruthenium-catalysed asymmetric hydrogenation reactions. The hydrogenation is run in $[bmim]PF_6$, and the products can conveniently be extracted from the IL by $scCO_2$. The extract is not contaminated with catalyst or IL, and the IL/catalyst solution can be reused several times without significant loss of activity.

Based on the same concept, Boesmann et al. [50] have developed a continuous-flow reactor system. For the hydrovinylation of styrene as a typical example, the catalyst was immobilised by dissolution in an IL and the reaction product was easily separated by extraction with $scCO_2$. The combination of IL with CO_2 provides the potential to activate and fine-tune the catalyst, while the whole process, being similar to a classical fixed-bed reactor, qualifies for large-scale industrial processes.

Catalysis in a biphasic IL/$scCO_2$ system has been demonstrated by Liu et al. [51]. For the hydrogenation of CO_2 in the presence of dialkylamines, the catalyst and the polar reaction intermediates dissolve predominantly in the IL phase, which results in high selectivity of the reaction. Easy catalyst recycling and product recovery is achieved by the fact that the reaction product dissolves predominantly in the CO_2 phase, while the catalyst stays immobilised in the IL.

9 Selected Applications in Homogeneous Catalysis

After having summed up the basic concepts that are preferentially used in combination with IL solvent chemistry, some prominent examples of successful applications from the recent chemical literature are highlighted in the following. The choice is necessarily quite subjective and by no means complete or exhaustive.

9.1 Hydrogenation and Hydroformylation

Hydrogenation and hydroformylation reactions are among the very first that were conducted in ILs [8]. Obviously, this is due to the great importance of these reaction classes for industrial processes. The majority of the reactions are fine examples of classical heterogeneous catalysis with all its advantages and disadvantages. Therefore, there is also a strong desire to combine the benefits of homogeneous and heterogeneous catalysis (Sect. 8). As an additional feature, Guernik et al. [52] have found that air-sensitive transition metal catalysts, such as Rh-MeDuPHOS, were stabler when dissolved in ILs. By using [bmim]PF_6 as the solvent, no protective atmosphere was needed in their asymmetric hydrogenation reactions. The reaction product was easily decanted after the reaction and the IL/catalyst solution was reused several times without significant loss of activation. As one of our own examples, we were able to demonstrate a similar effect for the enantioselective hydrogenation of trimethylindolenine in various ILs [27]. In this case, the reaction was not completely insensitive to air when run in an IL but significantly more so compared with when toluene is used. We attribute this effect to the significantly higher viscosities of our IL systems.

Berger et al. [53] have demonstrated that the solubility of molecular hydrogen differs substantially with different ILs. As an example, the solubility of H_2 in [bmim]BF_4 is almost 4 times higher than in [bmim]PF_6. Therefore, the authors point out that in these reactions the important kinetic parameter to be considered is the hydrogen *concentration* in the ionic phase (rather than the hydrogen *pressure*).

The use of ILs can also be superior to that of classical organic solvents in the hydrogenation of arenes. In present industrial applications, this reaction is performed in aqueous/organic biphasic systems [54], precluding the use of water-sensitive catalysts. The use of a biphasic IL/organic solvent system can overcome these problems [55]. Additionally, a new arene hydrogenation catalyst was found [56] that is highly active in ILs and with which hydrogenation of the arene ring of allylbenzene without hydrogenating the alkene bond is possible (Scheme 8).

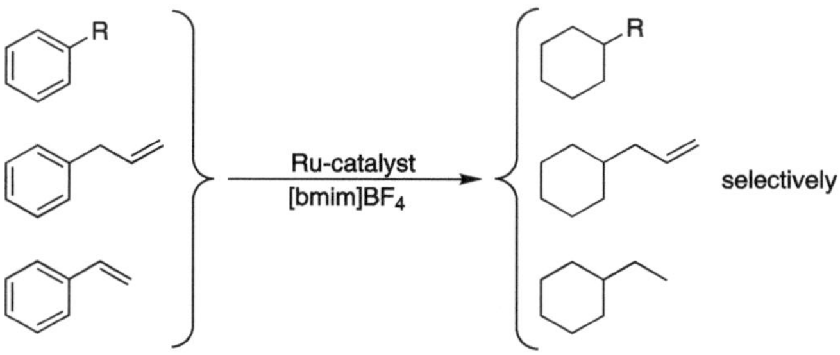

Scheme 8 Arene hydrogenation in ILs [55, 56]

Even the enantioselective hydrogenation of β-keto esters proceeds smoothly in various IL solvents. Ngo et al. [57] have reported conversions and enantiomeric excess values up to 99%, most of then being at least 97%.

For biphasic hydroformylation reactions, Brasse et al. [58] have introduced a new cobaltocenium-based bidentate ligand that shows high n-selectivity combined with all the "standard" advantages of IL-phase chemistry, like high catalyst activity and heterogenisation of a homogeneously catalysed reaction.

Favre et al. [59] have strikingly demonstrated how ILs can be optimised towards the reaction. For the biphasic rhodium-catalysed hydroformylation of 1-hexene, variation and optimisation of IL and ligand led, for example, to an improvement in turnover frequency by a factor of 7.

9.2 Palladium-Catalysed C – C-Coupling Reactions

A lot of interest has been focused on palladium-catalysed C – C bond-forming reactions owing to their importance for organic synthesis. Chen et al. [60, 61] have demonstrated that allylic alkylation reactions can readily be run in [bmim]BF_4 with all the common advantages (e.g., catalyst separation and recycling, as discussed earlier). The reaction is faster than in tetrahydrofuran and a wider choice of ligands is available. The authors claim that the allylpalladium intermediate is different in the two solvents (IL versus tetrahydrofuran), and that the results point to the existence of a "general IL effect".

The still very active field of Heck chemistry research has also been applied to the ionic phase. Herrmann and Boehm [62, 63] have exploited the use of their well-known phosphapalladacycles (Scheme 9, left) in various ILs (or NAILs as they call them, an acronym for non-aqueous ILs) based on imidazolium and ammonium cations. For their catalyst systems, [NBu_4]Br turned out to be the best choice. The catalysts showed improved efficiency, stability, and lifetime, without the need to add further promoting salt additives compared with conventional organic solvents like dimethylformamide. Catalyst recycling was successful for more than thirteen times without significant loss of activity. The same IL was used by Selvakumar et al. [64] with monocar-

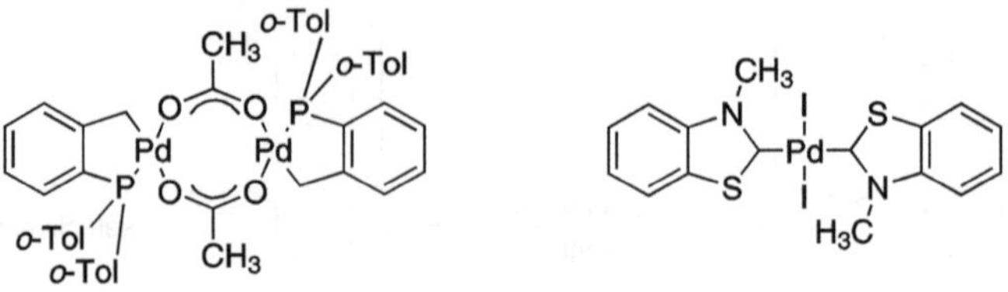

Scheme 9 Palladium catalysts for efficient Heck reactions in $[NBu_4]Br$. *Left*: Herrmann et al. [62, 63]. *Right*: Calò et al. [67–69]

benepalladium(0) complexes that were suitable catalysts for Heck reactions of aryl chlorides.

Xiao et al. [65] were able to demonstrate that a 2,2′-bisimidazole-based IL can act both as the solvent and as the ligand for catalytic reactions. Addition of $PdCl_2$ to this IL gave a catalytically active solution that was active in Heck coupling reactions.

Acceleration of Heck reactions in $[bmim]PF_6$ under microwave irradiation has been reported by Vallin et al. [66]. Owing to their polar nature, ILs are naturally perfect solvents for microwave irradiation. The coupling reactions reported were very fast and proceeded within 5–45 min.

Calò et al. [67, 68] have intensively studied the use of a palladium benzothiazole carbene complex (Scheme 9, right) for phosphine-free Heck reactions in $[NBu_4]Br$. They reported extremely high reaction rates, although they stated that they could not explain this high activity at the present date. The same catalyst/IL system was used for efficient arylation of α-substituted acrylates [69].

Arylation of an electron-rich olefin has also been reported by Xu et al. [70], using aryl iodides and bromides, and $Pd(OAc)_2$ in $[bmim]BF_4$. In this reaction, a regioselectivity greater than 99% was obtained. The results seem to demonstrate yet again a "general IL effect", which in this case, so the authors reason, is probably due to the acceleration of the ionic reaction pathway by the IL.

Among many other successful examples of palladium chemistry in ILs are the Stille coupling [71] and a copper-free version of the Sonogashira coupling [72] which has additionally been applied to a microflow system for efficient catalyst recycling. Revell and Ganesan [73] have demonstrated that the rate-accelerating IL effects can also be transferred to solid-phase conditions. The Suzuki–Miyaura cross-coupling of resin-bound iodophenol with various boronic acids was significantly accelerated by $[bmim]BF_4$.

9.3 Oxidation Reactions

ILs are, of course, suitable solvents for oxidation reactions, such as the oxidation of aromatic aldehydes with molecular oxygen and a nickel(II) catalyst [74]. For a TEMPO-CuCl catalysed aerobic oxidation of alcohols

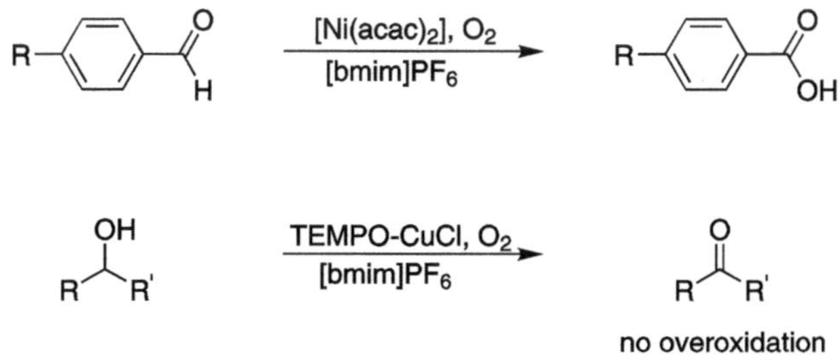

Scheme 10 Oxidation reactions in ILs [74, 75]

to aldehydes/ketones [75], it has been shown that the reaction, run in [bmim]PF_6, yielded no overoxidised carboxylic acid products (Scheme 10). The catalyst/IL solution could be reused for different types of substrates without any contamination of the earlier product. Alkene epoxidation with hydrogen peroxide and iron(III) porphyrins also ran smoothly, even in the halide-containing and thus easily and cleanly available IL [bmim]Br [76].

A speciality of ILs, namely the immobilisation of OsO_4 and its utilisation for olefin dihydroxylation, was published almost simultaneously by three groups. Yanada and Takemoto [77] have used [emim]BF_4 for the dihydroxylation of various olefins. They reported that the volatility of OsO_4 and therefore its toxicity is greatly suppressed when dissolved in the IL, which is clearly an advantage over the use of classical organic solvents. Yao [78] has used [bmim]PF_6 in combination with 4-(dimethylamino)pyridine to immobilise OsO_4 in the IL. Branco and Alfonso [79] have exploited biphasic and triphasic water/IL/*tert*-butanol systems for asymmetric olefin dihydroxylation with OsO_4. For a variety of simple olefins they report enantiomeric excesses up to 99% with full reusability of the catalyst.

9.4 Miscellaneous Reactions

For ruthenium-catalysed olefin metathesis reactions in conventional organic solvents, efficient and air-stable catalysts have been developed by the group of Grubbs [80, 81]. Nonetheless, these are not recyclable and are difficult to remove from the product. Buijsman et al. [82] have successfully performed ring-closing metathesis reactions in various ILs. In [bmim]PF_6 as the best solvent, full conversion was often achieved after 1-h reaction time with low ruthenium contamination of the product. In recycling experiments, the catalytically active IL solution showed a significant drop in activity as early as in the third run. Semeril at al. [83] have published a similar study using a ruthenium allenylidene catalyst. In their study, the solvent with the best results was [bmim]OTf. Apart from this detail, their results are very similar to those in the publication of Buijsman et al., even concerning the recycling experiments.

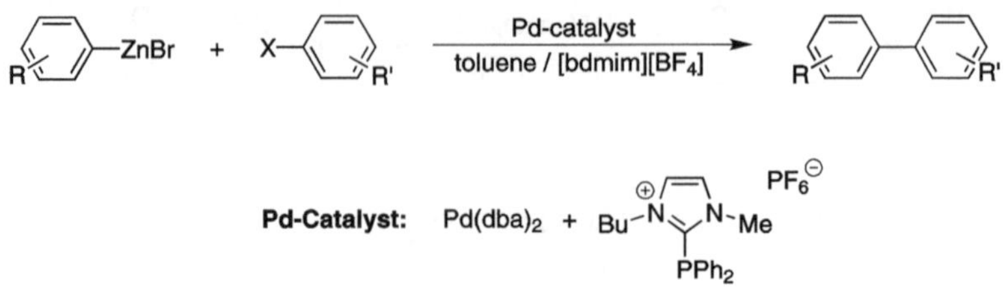

Scheme 11 Negishi cross-coupling in ILs using an imidazolium-based IL-derived phosphine ligand [84]

Sirieix et al. [84] have synthesised an ionic phosphine ligand that is derived from an imidazolium-based IL. They applied it to Negishi cross-coupling reactions in [bdmim]BF_4 (Scheme 11). By that, most reactions proceed at room temperature within a few minutes with yields around 70–92%. Catalyst separation after the reaction was easily accomplished, but recycling experiments again showed a drop in activity (20% lower yield, triple reaction time) in the third run.

Among many other successful applications of ILs in transition metal catalysis, there are also studies of Cr(salen)-catalysed asymmetric ring-opening reactions of epoxides [85], where a strong selectivity dependence on the nature of the anion in the IL was observed, and the first nickel(0)-catalysed coupling of aryl halides [86], which was done in [bmim]PF_6. Finally, even organocatalysis has found its way into the IL phase. Gou et al. [87] reported a considerable improvement in comparison with molecular solvents for an asymmetric direct aldol reaction— enantioselectivities of 91–99%, and the catalyst could be reused. The catalyst in this case was an *l*-prolinamide derivative.

10
Conclusion

In conclusion, I hope that I was able to demonstrate the usefulness of ILs for transition metal catalysis. The unique properties of these reaction media, in combination with their tunability, can lead to exceptionally high reactivities, selectivities, yields, and conversions, once the "right" IL is found for a given purpose. Their high polarity, in combination with variable miscibility with organic solvents, and their non-volatile nature give rise to easy catalyst heterogenisation and recycling techniques. And since an ever growing variety of ILs is now commercially available, it should be advantageous for everyone to get started using ILs as modern solvent systems.

Acknowledgement I would like to thank Deutsche Forschungsgemeinschaft for a generous Emmy Noether Fellowship.

References

1. Clark JH (1999) Green Chem 1:1
2. Betzemeier B, Knochel P (1999) Top Curr Chem 206:61
3. Leitner W (1999) Top Curr Chem 206:107
4. Dupont J, de Souza RF, Suarez PAZ (2002) Chem Rev 102:3667
5. Olivier-Bourbigou H, Magna L (2002) J Mol Catal A 182–183:419
6. Sheldon R (2001) Chem Commun 2399
7. Gordon CM (2001) Appl Catal A 222:101
8. Wasserscheid P, Keim W (2000) Angew Chem Int Ed Engl 39:3772
9. Welton T (1999) Chem Rev 99:2071
10. Holbrey JD, Seddon KR (1999) Clean Prod Proc 1:223
11. Davis JH Jr, Fox PA (2003) Chem Commun 1209
12. Wasserscheid P, Welton T (eds) (2003) Ionic liquids in synthesis, 1st edn. Wiley, Weinheim
13. Chauvin Y, Gilbert B, Guibard I (1990) Chem Commun 1715
14. Wilkes JS, Zaworotko MJ (1992) Chem Commun:965
15. Sugden S, Wilkins H (1929) J Chem Soc:1291 and references cited therein
16. Appleby D, Hussey CL, Seddon KR, Turp JE (1986) Nature 323:614
17. Kreher UP, Rosamilia AE, Raston CL, Scott JL, Strauss CR (2004) Molecules 9:387
18. Welton T (2003) In: Wasserscheid P, Welton T (eds) Ionic liquids in synthesis. Wiley, Weinheim, Sect 3.5
19. Reichardt C (2005) Green Chem 7:339
20. Holbrey JD, Rogers RD (2003) In: Wasserscheid P, Welton T (eds) Ionic liquids in synthesis. Wiley, Weinheim, Sect 3.1
21. Mantz RA, Trulove PC (2003) In: Wasserscheid P, Welton T (eds) Ionic liquids in synthesis. Wiley, Weinheim, Sect 3.2
22. Aki SNVK, Brennecke JF, Samanta A (2001) Chem Commun 413
23. Armstrong DW, He L, Liu Y-S (1999) Anal Chem 71:3873
24. Mathews CJ, Smith PJ, Welton T (2000) Chem Commun 1249
25. Davis JH Jr, Gordon CM, Hilgers C, Wasserscheid P (2003) In: Wasserscheid P, Welton T (eds) Ionic liquids in synthesis. Wiley, Weinheim, Chap 2
26. Cammarata L, Kazarian SG, Salter PA, Welton T (2001) Phys Chem Chem Phys 3:5192
27. Giernoth R, Krumm MS (2004) Adv Synth Catal 346:989
28. Holbrey JD, Seddon KR (1999) J Chem Soc Dalton Trans:2133
29. Gallo V, Mastrorilli P, Nobile CF, Romanazzi G, Suranna GP (2002) J Chem Soc Dalton Trans:4339
30. Klingshirn MA, Broker GA, Holbrey JD, Shaughnessy KH, Rogers RD (2002) Chem Commun 1394
31. Daguenet C, Dyson PJ (2004) Organometallics 23:6080
32. Doyle MP, Whitefleet JL, Bosch RJ (1979) J Org Chem 44:2923
33. Metcalf BW, Burkhart JP, Jund K (1980) Tetrahedron Lett 21:35
34. Hilgers C, Wasserscheid P (2003) In: Wasserscheid P, Welton T (eds) Ionic liquids in synthesis. Wiley, Weinheim, p 27
35. http://www.solvent-innovation.com
36. Knochel P (ed) (1999) Top Curr Chem 206
37. Carmichael AJ, Earle MJ, Holbrey JD, McCormac PB, Seddon KR (1999) Org Lett 1:997
38. Herrmann WA, Elison M, Fischer J, Kocher C, Artus GRJ (1995) Angew Chem Int Ed Engl 34:2371

39. Xu L, Chen W, Xiao J (2000) Organometallics 19:1123
40. Mathews CJ, Smith PJ, Welton T, White AJP, Williams DJ (2001) Organometallics 20:3848
41. McLachlan F, Mathews CJ, Smith PJ, Welton T (2003) Organometallics 22:5350
42. McGuinness DS, Mueller W, Wasserscheid P, Cavell KJ, Skelton BW, White AH, Englert U (2002) Organometallics 21:175
43. Aggarwal VK, Emme I, Mereu A (2002) Chem Commun 1612
44. Brown RJC, Dyson PJ, Ellis DJ, Welton T (2001) Chem Commun 1862
45. Mehnert CP, Cook RA, Dispenziere NC, Afeworki M (2002) J Am Chem Soc 124:12932
46. Mehnert CP (2005) Chem Eur J 11:50 and references cited therein
47. Mehnert CP, Mozeleski EJ, Cook RA (2002) Chem Commun 3010
48. Riisager A, Fehrmann R, Flicker S, van Hal R, Haumann M, Wasserscheid P (2005) Angew Chem 117:826; Angew Chem Int Ed Engl 44:815
49. Brown RA, Pollet P, McKoon E, Eckert CA, Liotta CL, Jessop PG (2001) J Am Chem Soc 123:1254
50. Boesmann A, Francio G, Janssen E, Solinas M, Leitner W, Wasserscheid P (2001) Angew Chem Int Ed Engl 40:2697
51. Liu F, Abrams MB, Baker RT, Tumas W (2001) Chem Commun 433
52. Guernik S, Wolfson A, Herskowitz M, Greenspoon N, Geresh S (2001) Chem Commun 2314
53. Berger A, de Souza RF, Delgado MR, Dupont J (2001) Tetrahedron Asymmetry 12:1825
54. Chauvin Y, Mussmann L, Olivier H (1995) Angew Chem Int Ed Engl 34:2698
55. Dyson PJ, Ellis DJ, Welton T, Parker DG (1999) Chem Commun 25
56. Boxwell CJ, Dyson PJ, Ellis DJ, Welton T (2002) J Am Chem Soc 124:9334
57. Ngo HL, Hu A, Lin W (2003) Chem Commun 1912
58. Brasse CC, Englert U, Salzer A, Waffenschmidt H, Wasserscheid P (2000) Organometallics 19:3818
59. Favre F, Olivier-Bourbigou H, Commereuc D, Saussine L (2001) Chem Commun 1360
60. Chen W, Xu L, Chatterton C, Xiao J (1999) Chem Commun 1247
61. Ross J, Chen W, Xu L, Xiao J (2001) Organometallics 20:138
62. Herrmann WA, Boehm VPW (1999) J Organomet Chem 572:141
63. Boehm VPW, Herrmann WA (2000) Chem Eur J 6:1017
64. Selvakumar K, Zapf A, Beller M (2002) Org Lett 4:3031
65. Xiao JC, Twamley B, Shreeve JM (2004) Org Lett 6:3845
66. Vallin KSA, Emilsson P, Larhed M, Hallberg A (2002) J Org Chem 67:6243
67. Calò V, Nacci A, Lopez L, Mannarini N (2000) Tetrahedron Lett 41:8973
68. Calò V, Nacci A, Monopoli A, Lopez L, di Cosmo A (2001) Tetrahedron 57:6071
69. Calò V, Nacci A, Lopez L, Napola A (2001) Tetrahedron Lett 42:4701
70. Xu L, Chen W, Ross J, Xiao J (2001) Org Lett 3:295
71. Handy ST, Zhang X (2001) Org Lett 3:233
72. Fukuyama T, Shinmen M, Nishitani S, Sato M, Ryu I (2002) Org Lett 4:1691
73. Revell JD, Ganesan A (2002) Org Lett 4:3071
74. Howarth J (2000) Tetrahedron Lett 41:6627
75. Ansari IA, Gree R (2002) Org Lett 4:1507
76. Srinivas KA, Kumar A, Chauhan SMS (2002) Chem Commun 2456
77. Yanada R, Takemoto Y (2002) Tetrahedron Lett 43:6849
78. Yao Q (2002) Org Lett 4:2197
79. Branco LC, Afonso CAM (2002) Chem Commun 3036
80. Nguyen ST, Grubbs RH (1993) J Am Chem Soc 115:9858

81. Scholl M, Ding S, Lee CW, Grubbs RH (1999) Org Lett 1:953
82. Buijsman RC, van Vuuren E, Sterrenburg JG (2001) Org Lett 3:3785
83. Semeril D, Olivier-Bourbigou H, Bruneau C, Dixneuf PH (2002) Chem Commun 146
84. Sirieix J, Ossberger M, Betzemeier B, Knochel P (2000) Synlett 1613
85. Song CE, Oh CR, Roh EJ, Choo DJ (2000) Chem Commun 1743
86. Howarth J, James P, Dai J (2000) Tetrahedron Lett 41:10319
87. Gou HM, Cun LF, Gong LZ, Mi AQ, Jiang YZ (2005) Chem Commun 1450

Top Curr Chem (2007) 276: 25–68
DOI 10.1007/128_064

Published online: 29 July 2006

Transfer of Parahydrogen-Induced Hyperpolarization to Heteronuclei

Lars T. Kuhn (✉) · Joachim Bargon

Institute of Physical & Theoretical Chemistry, University of Bonn, Wegelerstrasse 12, 53115 Bonn, Germany
lars.kuhn@chem.ox.ac.uk

Abstract Homogeneously catalyzed hydrogenation reactions of unsaturated substrates with H_2 gas mixtures enriched in parahydrogen yield strong NMR signal enhancements of the transferred ^{1}H nuclei if the symmetry of H_2 is broken in the resulting hydrogenated products. This chemically induced hyperpolarization phenomenon known as parahydrogen-induced polarization (PHIP) is a well-established polarization technique in

NMR spectroscopy. Ever since its theoretical prediction and subsequent experimental verification the method has been used to increase signal intensity in ^{1}H-NMR spectroscopy for the elucidation of catalytic pathways of hydrogenation reactions and their kinetic behavior. Furthermore, PHIP is not confined to the attached protons and ^{1}H nuclei which are close to the hydrogenation site but it can also be transferred spontaneously to heteronuclei, which are present in the hydrogenation product. In this review we give an overview of the different experiments that have been performed in recent years in order to efficiently transfer PHIP-derived polarization to insensitive magnetically active nuclei following the catalyzed parahydrogenation of their unsaturated precursor molecules. A detailed description of the experiments dealing with every individual heteronucleus in particular is followed by a discussion of the mechanisms leading to PHIP transfer. Subsequently, we describe the existing set of pulse sequences that have been designed and successfully employed in order to induce an exchange of increased magnetization originating from PHIP between protons and heteronuclei using conventional coherence transfer schemes. Finally, possible applications of non-proton PHIP spectroscopy in medicine and clinical research are outlined.

Keywords NMR · PHIP · Parahydrogenation · Homogeneous catalysis · Hyperpolarization transfer

Abbreviations

NMR	nuclear magnetic resonance
PHIP	parahydrogen-induced polarization
NOE	nuclear Overhauser effect
ALTADENA	adiabatic longitudinal transport after dissociation engenders nuclear alignment
PASADENA	parahydrogen and synthesis allow dramatically enhanced nuclear alignment
MRS	magnetic resonance spectroscopy
MRT	magnetic resonance tomography
INEPT	insensitive nuclei enhanced by polarization transfer
DEPT	distortionless enhancement by polarization transfer
INADEQUATE	incredible natural abundance double quantum transfer experiment
n.a.	natural abundance
PART	parahydrogen-aided resonance transfer
SE	signal enhancement
SEPP	selective excitation of polarization using PASADENA
CIDNP	chemically induced dynamic nuclear polarization
TOCSY	total correlation spectroscopy
FID	free induction decay
SNR	signal-to-noise ratio

1 Introduction

1.1 Parahydrogen-Induced Polarization (PHIP)

Nuclear magnetic resonance (NMR) is a well-established spectroscopic technique which suffers inherently from a rather low sensitivity. However, there are numerous physical and a number of chemical ways to circumvent this sensitivity problem. Most of the attempts to increase NMR-related sensitivity in a physical way rely on a modification of the field strengths of NMR spectrometers or the use of specially designed cryoprobes. NMR sensitivity can also be increased using pulse sequences which transfer magnetization from sensitive to rather insensitive magnetically active nuclei. Two chemical ways to increase NMR sensitivity are Chemically Induced Dynamic Nuclear Polarization (CIDNP) and Parahydrogen-Induced Polarization (PHIP). The former relies on the intermediate generation of free radicals whose recombination rate depends on the spin states of NMR-active nuclei which are present in the interacting paramagnetic substrates. The other method (PHIP) uses a completely different approach employing the hydrogenation of unsaturated precursor molecules with parahydrogen, a spin isomer of molecular dihydrogen (H_2). In this review, we will focus on the latter technique and we will outline how PHIP has been employed in recent years to selectively polarize heteronuclei of a large number of structurally different substrate molecules.

Hydrogenation reactions using H_2 (dihydrogen) play a significant role in organic synthesis. Typically, these reactions have to be catalyzed in order to activate the H_2 molecule and to lower the activation energy of the subsequent hydrogen transfer to a substrate molecule using either heterogeneous or homogeneous catalysts. About two decades ago, it was found that homogeneously catalyzed hydrogenation reactions of unsaturated substrates with H_2 enriched in parahydrogen lead to the observation of strongly enhanced absorptive and emissive signals of the two transferred protons in the resulting ^{1}H-NMR spectra of the hydrogenation products. The observation of this extraordinary phenomenon was theoretically predicted by Bowers et al. in 1986 [1]. Shortly afterwards Weitekamp et al. [2] and Eisenschmid et al. [3] could prove experimentally that the phenomenon actually occurs when certain experimental requirements are met. The effect, which has been extensively studied ever since its theoretical prediction and experimental verification, has been termed PASADENA (Parahydrogen And Synthesis Allow Dramatically Enhanced Nuclear Alignment) [2] or PHIP (Parahydrogen-Induced Polarization) [3] and originates from the breaking of the high symmetry of the dihydrogen molecule present in its *para* spin state ($I = 0$) during the course of the hydrogenation reaction. The characteristic signal patterns in the resulting NMR spectra are associated with a signal enhancement (SE) of several orders of magnitude (typically up to four). The PHIP phenomenon, however, only occurs if the two hydrogen atoms of parahydrogen are transferred jointly to the unsaturated center of the substrate. The simultaneous transfer of the two parahydrogen atoms is crucial for the observation of PHIP signals since it makes sure that the original spin correlation, i.e., a resulting singlet in the case of parahydrogen with $I = 0$ and a triplet for orthohydrogen with $I = 1$, between these two nuclei is maintained throughout the whole hydrogenation cycle and also afterwards in the products. Moreover, the patterns of the polarization signals depend strongly on the way every PHIP experiment is conducted [4, 5].

Individual PHIP spectra are normally acquired in situ, i.e., immediately while hydrogenating unsaturated substrates either within low or high magnetic fields. Both experimental setups differ in that the hydrogenation process is either initiated in the presence of just the Earth's magnetic field (0.05 mT), i.e., outside the strong field of the NMR magnet, followed by an immediate transfer of the NMR tube to the spectrometer and the acquisition of the polarization spectrum (ALTADENA condition; ALTADENA: Adiabatic Longitudinal Transfer After Dissociation Engenders Nuclear Alignment) [6]. This setup eventually leads to the observation of NMR signals exhibiting *net* polarization in the resulting spectra showing either enhanced absorption *or* emissionb (Fig. 1). Alternatively, the whole experiment can be conducted with the NMR sample being inside the sensitive coil region of the spectrometer through-

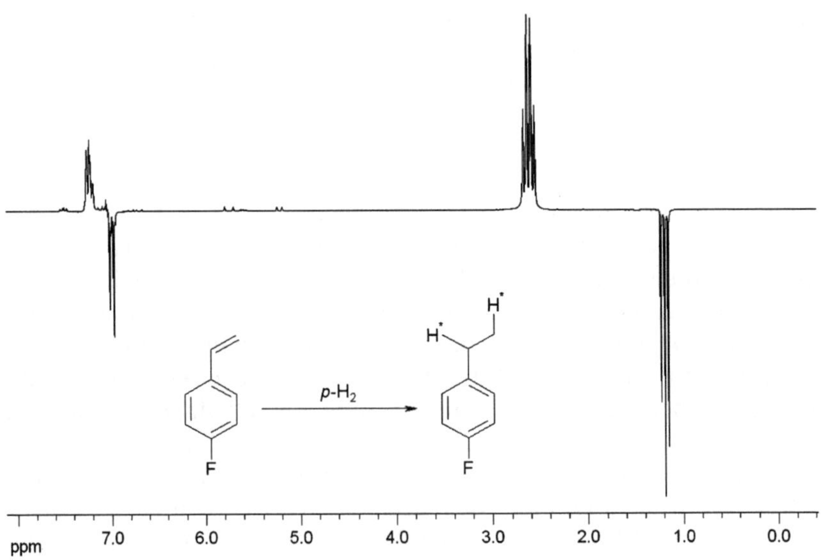

Fig. 1 Single scan ^{1}H-ALTADENA spectrum acquired during the parahydrogenation of 4-fluorostyrene yielding 4-fluoroethylbenzene. Apart from the two transferred protons also the aromatic 2,6- (E) and 3,5- (A) protons show strong polarization due to homonuclear dipolar cross-relaxation

out the whole experiment (PASADENA condition; PASADENA: Parahydrogen And Synthesis Allow Dramatically Enhanced Nuclear Alignment) leading to characteristic *antiphase* patterns of the polarization signals obtained.

Due to the significant sensitivity enhancement associated with this method, PHIP has been proven to be a powerful tool for the in situ investigation of homogeneously catalyzed hydrogenation reactions providing information about the fate of the H_2 molecule, the catalyst, and of the substrate using NMR spectroscopy.

1.2 Transfer of PHIP-Derived Polarization to Heteronuclei

Ever since the discovery of the PHIP phenomenon, scientists have been interested not only in the generation of ^{1}H-polarization of the two protons stemming from the parahydrogen molecule itself, but also in transferring the high spin order of the parahydrogen molecule to insensitive heteronuclei of either the dihydride intermediate formed during the catalytic hydrogenation cycle or even to heteronuclei present in the reaction product.

The first observation of a transfer of polarization to a heteronucleus during a PHIP experiment was reported by Eisenschmid et al. in 1989 [7]. These authors were able to show that PHIP-derived polarization could be transferred to the hydrogenation catalyst, namely to the two chemically inequivalent ^{31}P nuclei of the diphenylphosphinoethane ligand of a planar Iridium(I) complex during its oxidative addition using parahydrogen-enriched H_2 in the absence of the strong magnetic field of the NMR spectrometer (ALTADENA). The two polarized heteronuclei exhibited a characteristic antiphase signal pattern in the ^{31}P-NMR spectrum which was acquired immediately after the addition of a parahydrogen-enriched gas mixture to the sample solution inside the NMR tube. In their publication the authors mentioned that the spectrum acquired in this "in situ" way was very similar to a spectrum they could also attain when an INEPT experiment was applied to the same solution of reactants treated with "regular" H_2, i.e., thermally equilibrated dihydrogen gas. Furthermore, the authors concluded that dipolar interactions were responsible for the transfer of PHIP to the two ^{31}P nuclei.

Several years later, Duckett et al. studied the same reaction, again using parahydrogen as the oxidizing agent of the Iridium(I) complex [8]. This time, however, the observed nucleus was ^{13}C and the authors could observe a signal enhancement (SE) of 95 for the ^{13}C nucleus of the carbonyl ligand of the catalyst compared to the same reaction using thermally equilibrated dihydrogen gas. The SE observed for the ^{13}C nucleus was substantial even when the unlabeled complex was parahydrogenated. This phenomenon was named PART (Parahydrogen-Aided Resonance Transfer) accordingly.

The first observation of PHIP transfer to the magnetically active heteronucleus of a hydrogenation product was achieved by Bargon and co-workers in 1995 during the *cis*-parahydrogenation of acetylenedicarboxylic dimethyl ester yielding maleic acid dimethyl ester [9]. In this case, the polarization stemming from the parahydrogen molecule could be transferred to both the chemically equivalent ^{13}C nuclei of the double bond and also to those of the carbonyl group and the methyl group of the hydrogenation product maleic acid dimethyl ester.

The observation of this PHIP-derived transfer was remarkable for two reasons. First, their single scan in situ experiment yielded a SE factor for the carbonyl ^{13}C nuclei of 2580 and was thus more than substantial. Second, the observation of any PHIP transfer within this system was completely unexpected since the hydrogenation of the starting material yielded a symmetric product which, according to the status of the theory of PHIP at that time, should not exhibit any polarization signals in its NMR spectrum at all. The authors, however, could show that the presence of naturally abundant ^{13}C was responsible and sufficient for the breaking of the molecule's magnetic symmetry thereby allowing the appearance of PHIP in the parahydrogenated products. The very efficient transfer of PHIP in this case could be explained with the fact that due to the symmetric nature of the substrate molecule the two attached protons of the hydrogenation product form essentially a strongly coupled AA′X *three-spin system* together with every individual ^{13}C nucleus, which makes the polarization transfer extremely favorable.

In all these aforementioned systems PHIP transfer was exclusively observed when the parahydrogenation reaction was carried out in low magnetic fields using ALTADENA conditions. In 1996, however, Bargon and co-workers could demonstrate that the transfer of polarization from parahydrogen to heteronuclei does not just occur *spontaneously* as described before, but that it can also be induced at high magnetic fields provided that specifically designed pulse sequences are applied during or immediately after the hydrogenation reaction [10–13]. They devised a number of DEPT and INEPT experiments including conventional pulse schemes to efficiently transfer coherence stemming from p-H_2 to ^{13}C, ^{15}N, and ^{29}Si nuclei. A SE of ca. 500 was consistently attained in this case. Additionally, a parahydrogen variant of the very powerful INADEQUATE experiment (PH-INADEQUATE) was devised in the same year by Natterer et al. which could overcome the sensitivity problems of the normal INADEQUATE experiment in a very elegant way using increased ^{13}C polarization originating from the p-H_2 molecule.

A more comprehensive analysis of ^{13}C-PHIP polarization spectra was carried out by Bargon and co-workers in 2002 using a set of structurally similar aliphatic (1-hexyne, 2-hexyne, 3-hexyne, 4-octyne, and 3,3-dimethyl-1-butyne) and a single aromatic alkyne (phenylacetylene) derivative [14]. In all cases PHIP was transferred efficiently to every nucleus of the carbon skeleton of every individual molecule using an ALTADENA setup. Furthermore, the authors could name certain steric and electronic requirements which have to be met for an efficient polarization transfer to occur. As far as the transfer mechanisms were concerned, however, it was very difficult to elucidate the details of the individual PART phenomena.

In more recent studies, we have examined the spontaneous transfer of PHIP to ^{19}F using a set of structurally similar singly fluorinated styrene and phenylacetylene derivatives [15]. We were able to show that the unprecedented and very efficient transfer of PHIP-derived polarization to ^{19}F occurs using either ALTADENA (low field) or PASADENA (high field) conditions. Furthermore, we were able to shed more light on the mechanisms which govern PART to heteronuclei by showing that dipolar as well as scalar couplings mediate this transfer, the details depending on whether high-field or low-field conditions are applied during the hydrogenation reaction.

Today, the observation of PHIP transfer is a routinely used procedure to increase the sensitivity of NMR experiments. A large number of mostly insensitive heteronuclei, reaching from ^{2}H, ^{13}C, ^{15}N, ^{29}Si, and ^{31}P to ^{19}F, are able to benefit from this transfer of the parahydrogen-derived high spin order showing a significant signal enhancement in the resulting NMR spectra of the parahydrogenation products. In almost all cases, however, PHIP transfer to heteronuclei was almost exclusively observed under ALTADENA conditions, i.e., when the parahydrogenation reaction was initiated in the Earth's magnetic field followed by an immediate and adiabatic transfer of the sample to the NMR spectrometer and the subsequent acquisition of the spectrum. Also, the mechanisms which govern PHIP transfer are still far from being understood in full detail.

In the following, we will give a detailed description of the observation of PHIP-derived polarization transfer to heteronuclei dealing with every individual nucleus separately. We will also present recent results which are intended to clarify the elucidation of the mechanisms underlying the PART phenomenon. Subsequently, we describe in detail all pulse sequences which have been devised so far to induce a polarization transfer stemming from PHIP. The survey is concluded by a description of applications of parahydrogen-derived hyperpolarization of heteronuclei ("hetero PHIP") in medicine and clinical research, and a brief outlook into future perspectives of the method is given.

2 PHIP Transfer to Individual Heteronuclei

Homogeneously catalyzed hydrogenation reactions with parahydrogen leading to strong ^{1}H-polarization signals in the respective NMR spectra also give rise to strong heteronuclear polarization, especially when the hydrogenations are carried out in low magnetic fields. As a typical example, the polarization transfer from ^{1}H to ^{13}C nuclei during the parahydrogenation of alkynes has been investigated using several different substrate molecules. It could be shown that in systems containing easily accessible triple bonds, e.g., phenylethyne or 2,2-dimethylbutyne, a polarization transfer to virtually all carbon nuclei of the molecule occurs. Accordingly, all ^{13}C resonances can be observed in the NMR spectra of the hydrogenation product recorded in situ with a good to excellent signal-to-noise ratio (SNR) using only a single transient. This technique thus permits the structural elucidation of a large number of organic compounds exploiting the PHIP effect. In the following section the qualitative influence of different substituents on the symmetry and the electronic structure of the substrate and its parahydrogenation product are discussed together with their influence on the efficiency or the feasibility of a transfer of polarization to heteronuclei. Furthermore, spectroscopic evidence has been gained for an initial attachment especially of hydrogenation products containing aromatic segments to the transition metal center of the cationic hydrogenation catalyst usually employed for a PHIP experiment as the activating species.

It has been postulated and demonstrated before that "hetero PHIP" for nuclei like ^{13}C, ^{15}N, ^{29}Si, and ^{31}P can result in a signal enhancement (SE) of more than two orders of magnitude if the reactions are carried out in low magnetic fields using ALTADENA conditions. Barkemeyer et al. [9] and Natterer et al. [12, 13] previously showed that strong enhancements can also be achieved in high magnetic fields when hydrogenating symmetric systems where the breakdown of the symmetry is caused by the naturally abundant ^{13}C nuclei occurring individually in the two other equivalent carbon atoms of the unsaturated double bond of the substrate.

Although this phenomenon provides a powerful tool for the NMR investigation of nuclei with a low sensitivity, the full potential of this sizeable polarization transfer from parahydrogen to other nuclei has not been applied very much as compared to ^{1}H PHIP NMR spectroscopy. Especially the use of the PHIP effect for the enhancement and correspondingly simplified detection of ^{13}C- or ^{15}N-nuclei is a useful tool for the structural investigation of organic molecules.

2.1 PHIP Transfer to ^{1}H

Transferring PHIP-derived polarization to magnetically active nuclei of the parahydrogenation product opens up a variety of attractive options. The most obvious and most common transfer of polarization from the two initially polarized parahydrogen nuclei is that to nearby protons through cross-relaxation caused by dipolar interactions. These processes are well known in magnetic resonance spectroscopy since they give rise to the nuclear Overhauser effect (NOE). The observation of this homonuclear polarization transfer is almost always observed during the parahydrogenation of suitable precursor molecules when ^1H nuclei are present in the immediate vicinity of the hydrogenation site and can thus appreciably interact with the two attached protons stemming from the parahydrogen molecule [4, 5, 16, 17]. A detailed description of the mechanisms dealing with this homonuclear PHIP transfer is given in Sect. 3 of this review.

Another approach to transfer polarization stemming from the former parahydrogen ^1H-nuclei to other protons of the same molecule or, alternatively, even to protons of nearby molecules exploiting an intermolecular ^1H–^1H polarization transfer was suggested by Hübler et al. in 2000 [18]. In an attempt to prove the existence of an intermediate substrate-catalyst complex, which is thought to be formed during the hydrogenation of a suitable precursor molecule, they devised two pulse sequences (PH-1D-NOESY, PH-1D-ROESY) which are able to selectively transfer magnetization from a hyperpolarized parahydrogen proton of the parahydrogenated substrate molecule to a nearby aromatic proton of the hydrogenation catalyst [1,4-bis(diphenylphosphino)butane]-(1,5-cyclooctadiene)-rhodium(I) exploiting selective excitation and subsequent dipolar cross-relaxation.

2.2 PHIP Transfer to ^2H

Furthermore, PHIP transfer could not just be detected or induced between protons but it has also recently been observed between the former parahydrogen nuclei and ^2H nuclei present in the hydrogenated product molecule during the parahydrogenation reactions of acetylene-d_2 to ethylene-d_2 and ethylene-d_4 to ethane-d_4 (Fig. 2) [19].

The ^2H-NMR spectra of these two parahydrogenated symmetrical molecules acquired using ALTADENA conditions showed enhanced antiphase signals and the splitting between the absorptive and the emissive peaks was equal to the direct coupling constant between the ^2H-nucleus and the ^{13}C-nucleus directly attached to it. The authors stated that this observation was an unambiguous indication that polarization transfer in this case is mediated

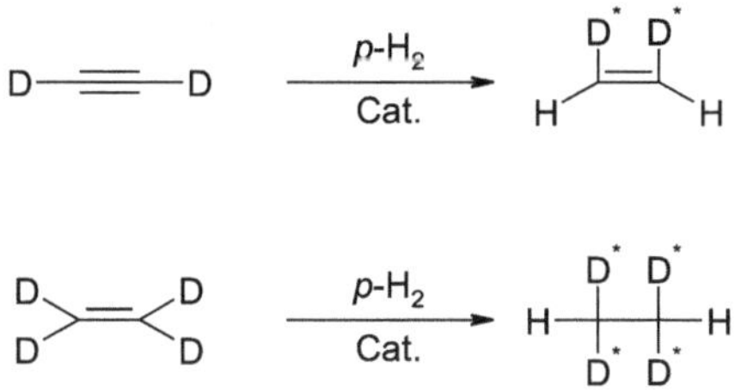

Fig. 2 Parahydrogenation reactions of deuterated precursor molecules conducted using ALTADENA conditions. The polarization stemming from the former p-H_2 molecule is efficiently transferred to all ^2H nuclei of the parahydrogenation product

by naturally abundant ^{13}C-nuclei and that the active product operator transforms into an antiphase signal upon application of a read pulse to deuterium. This finding created a new understanding of the transfer of PHIP to heteronuclei since the paper published by Aime et al. provided the first experimental evidence and also a subsequent theoretical description of a PHIP transfer to heteronuclei which did not involve a direct participation of the two attached ^{1}H nuclei stemming from the parahydrogen molecule itself.

2.3 PHIP Transfer to ^{13}C

The magnetically active nucleus ^{13}C is one of the most widely studied probes in NMR spectroscopy for the structural elucidation and characterization of organic molecules. Due to its low natural abundance and its low sensitivity, however, ^{13}C-NMR studies are rather difficult to accomplish, and experiments usually take up large amounts of spectrometer time in order to attain an appreciably high signal-to-noise ratio. It was discovered very early though that the sensitivity of ^{13}C is substantially increased when hyperpolarization originating from parahydrogen can be transferred to the ^{13}C-nucleus exploiting the PART phenomenon. Because of this, ^{13}C is today one of the most well studied "hetero PHIP" nuclei, and it has attracted wide attention among scientists dealing with the PHIP phenomenon.

The first observation of a Parahydrogen-Induced Polarization Transfer to ^{13}C was reported by Duckett et al. in 1993 [8]. The reaction chemistry employed in this study was the well-known oxidative addition of H_2 to the ^{13}C-labeled complexes $IrCl(^{13}CO)(PPh_3)_2$ and $IrBr(^{13}CO)(dppe)$ which proceeds in a concerted way (Fig. 3).

The utilization of para-enriched hydrogen gas in this case yielded the expected appearance of spin-polarized hydride NMR resonances of the oxidized transition metal complex upon placement of the sample into the NMR probe immediately after thawing from 77 K and shaking to dissolve the applied p-H_2 gas mixture. Furthermore, the authors were able to observe an efficient transfer of the parahydrogen polarization to the ^{13}C-nucleus of the carbonyl ligand of this complex applying a standard INEPT+ sequence [20] immediately after the transfer of the sample into the sensitive coil region of the spectrometer.

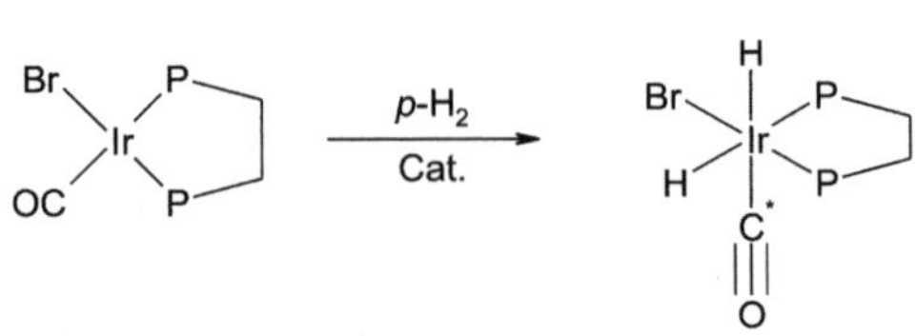

Fig. 3 Scheme of the parahydrogenation reaction during which the first PHIP-derived polarization transfer to a ^{13}C nucleus was observed

The enhancement factor attained for the carbon nucleus was very high and ^{13}C polarization of the carbonyl ligand could thus also be achieved when unlabeled samples were employed.

In 1995 Bargon and co-workers were able to observe the first transfer of PHIP to the ^{13}C-heteronuclei of a parahydrogenation product employing the hydrogenation of acetylenedicarboxylic dimethyl ester yielding maleic acid dimethyl ester (Fig. 4) [9]. In this case, the polarization stemming from p-H_2 was transferred to all ^{13}C-nuclei of the hydrogenation product and the amount of polarization was substantial. The signal enhancement (SE) factor for the carbonyl carbon nuclei of the methoxy group exceeded 2580 corresponding to a signal-to-noise ratio of 300. Obtaining an equivalent signal-to-noise ratio of the product resonances without using an enriched parahydrogen gas mixture would have required 6.7×10^6 transients and substantially more spectrometer time accordingly. An additional heterodecoupling of the methoxy protons yielded an increase of the signal-to-noise ratio of up to 800.

The observation of this polarization transfer made by Barkemeyer et al. was more than unexpected and represented a remarkable hallmark in the short history of PHIP transfer for several reasons: First, the attained enhancement factors for all ^{13}C nuclei of the hydrogenated product molecule exceeded four orders of magnitude consistently and were thus more than substantial. Second, the observation of any PHIP enhancement of magnetically active nuclei of the product molecule was completely unexpected due to the symmetric nature of both the starting compound and the hydrogenation product. In the latter, the jointly transferred protons can be found in chemically equivalent positions and since a breaking of the high symmetry of the parahydrogen molecule does thus not occur upon its addition to the substrate, the PHIP effect itself and also a PHIP transfer to any magnetically active nucleus of the product should normally not occur. The lucid explanation of their experimental findings, however, was very convincing and led to an important extension of the general theory of PHIP [21]. According to this concept, the two protons stemming from the parahydrogen molecule are present in chemically equivalent positions after their catalyzed addition to the substrate molecule. However, the presence of naturally abundant ^{13}C in just a minor fraction of all substrate molecules (n.a. (^{13}C) $\sim$ 1.1%) present in the reaction mixture breaks the magnetic symmetry of the molecule and this is sufficient for the observation of PHIP both on the protons and subsequently

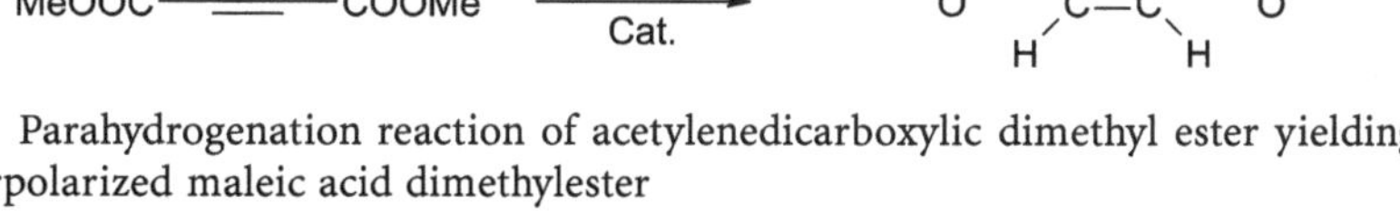

Fig. 4 Parahydrogenation reaction of acetylenedicarboxylic dimethyl ester yielding ^{13}C-hyperpolarized maleic acid dimethylester

also on the heteronuclei. Additionally, the very efficient transfer of PHIP in this case could be explained with the fact that due to the symmetric nature of the product the two attached protons of the hydrogenation product form a strongly coupled AA′X three-spin system together with any ^{13}C nucleus of the product, rendering an efficient polarization transfer extremely favorable.

A subsequent study and more detailed analysis of the origin of ^{13}C-PHIP was carried out by Stephan et al. employing a set of structurally and chemically similar alkyne derivatives following their in situ parahydrogenation (5 bar) with the cationic transition metal catalyst [1,4-bis(diphenylphosphino)butane]-(1,5-cyclooctadiene)rhodium(I) as the activating species and $CDCl_3$ or acetone-d_6 as the solvent [14]. The substrates to be hydrogenated were the aliphatic alkynes 3,3-dimethylbut-1-yne, 1-hexyne, 2-hexyne, 3-hexyne, 4-octyne, and the aromatic alkyne phenylethyne. These substrates were chosen for the following reasons: First, the triple bond of all substrates is rather accessible to the hydrogenation catalyst making an efficient hydrogenation likely to occur. Second, the substituents cover a wide range of different electronic structures and sterical requirements, and this fact allowed a rather systematic way of probing certain conditions that have to be met for a polarization transfer to occur. Also the substrates exhibit different geometric features, which enabled the investigators to probe the influence of the symmetric properties of these substrates on the efficiency of polarization transfer.

Comparing the selected hexyne derivatives one can see that their electronic structure and the accessibility of their individual triple bonds are very similar whereas the symmetry of the bond is very different. Moreover, 1-hexyne, 3,3-dimethylbutyne, and phenylethyne all have terminal triple bonds. However, the electronic structure and the accessibility of the π-system of these triple bonds are influenced differently by the activating hydrogenation catalyst depending on the electronic effect of the functional group adjacent to these triple bonds. Comparing the ^{13}C PHIP spectra of all selected compounds the authors were able to show how +M and +I effects together with the effect of a modulated sterical hindrance influence the hydrogenation and the polarization transfer from the protons to individual ^{13}C nuclei.

Furthermore, the authors compared the attained corresponding signal enhancements as well as the magnitude of the polarization transfer from the former parahydrogen nuclei to the directly adjacent former alkynic carbon atom and also to the other carbon atoms of the respective parahydrogenation products using low-field conditions (ALTADENA). In these experiments the hydrogenations were carried out in the presence of just the Earth's magnetic field followed by the transfer of the sample into the high magnetic field of the NMR spectrometer and the acquisition of the ^{13}C-PHIP spectra. By means of this procedure, the polarization was transferred to magnetically active heteronuclei in the following way: The resonance frequencies of all nuclei in the very low magnetic field of the earth are virtually the same. Thus, a coupled spin system of a high order is obtained, i.e., the difference of the resonance frequencies of ^{1}H- and ^{13}C-nuclei are small compared to their coupling constants. As will be shown in Section 3 this strong coupling is essential for an efficient polarization transfer from protons to a large number of carbon nuclei to occur. However, in the high field of a superconducting NMR magnet the polarization transfer to heteronuclei is less efficient because the difference in resonance frequencies between ^{1}H and every heteronucleus is significant and exceeds the magnitude of the coupling constants between the heteronuclei and the protons. In this latter case, polarization transfer can be achieved most effectively using appropriate pulse sequences (vide infra).

Even though "hetero PHIP" spectra are usually recorded using only a single transient they exhibit a very good to excellent signal-to-noise ratio (SNR). Because of the strong signal enhancement caused by the PHIP effect typically no signals are visible for the carbon atoms of the starting material, because in general it is not possible to obtain non-polarized ^{13}C spectra with an acceptable SNR using just a single transient. To appreciate or calibrate the high signal enhancement due to PHIP, the intensity of the NMR signal of the solvent $CDCl_3$ may serve as a measure: In a conventional NMR spectrum the residual resonance of the solvent has a similar or even larger intensity than the resonance of the starting material. Nonetheless, in the ^{13}C-PHIP spectra recorded as outlined here the signal stemming from $CDCl_3$ does not show up at all. Normally, the solvent may give rise to the strongest signal of the spectrum and thus has to be suppressed using appropriate pulse sequences prior to the acquisition of the spectrum. In the case of ^{13}C-PHIP, however, this is not necessary at all. Comparing both ^{1}H- and ^{13}C-PHIP spectra one can see that the latter exhibit much more pronounced differences in polarization intensity when different substrate molecules are employed. In addition, a homonuclear transfer from the former p-H_2 protons to other protons in the molecule is only observed for distances of up to three covalent bonds ($^{3}J_{HH}$) whereas in the case of ^{13}C PHIP a transfer of the polarization to carbon nuclei as far away as six covalent bonds has been detected.

It is worthwhile pointing out that in the study by Stephan et al. all spectra were acquired under identical conditions, e.g., hydrogen pressure, elapsed time prior to the acquisition of the spectrum, temperature, and concentration of the chemicals used. These spectra are thus considered to be characteristic for every individual system. Nonetheless, it would be desirable to use a setup, which permits the acquisition of ALTADENA "hetero PHIP" spectra in a totally standardized fashion to increase the reproducibility of the experiment itself. Such a system would allow the investigator to conduct the parahydrogenation reaction in a very low magnetic field followed by the adiabatic transfer of the sample into the NMR spectrometer for a subsequent spectroscopic analysis.

In addition to the transfer of polarization to all carbon nuclei of the hydrogenation products, a slight low-field shift of the ^{13}C-resonances in the aromatic region of the PHIP spectrum of the product molecules recorded during the hydrogenation of phenylethyne could be observed. A similar effect had been observed before in the ^{1}H-PHIP NMR spectra recorded during the hydrogenation of styrene derivatives using cationic Rh(I) catalysts and could be explained with the occurrence of a so-called "*product attachment*" during the catalytic cycle of the hydrogenation reaction [22, 23]. This term implies that the hydrogenation product binds to the metal center of the catalyst immediately after its formation via a π-complex between the aromatic system of the phenyl ring and one of the *d*-orbitals of the transition metal complex (Fig. 5). The strength of this attachment depends on the electronic structure of the product and the catalyst, and it either leads to a shift of resonances to higher or lower frequencies in the respective NMR spectrum. Due to the increased chemical shift dispersion of ^{13}C compared to ^{1}H, this effect is much more pronounced when observing the ^{13}C resonances ($\Delta\delta \sim 4$ ppm) of the attached product as compared to the shift in the ^{1}H spectrum ($\Delta\delta \sim 0.5$ ppm).

During the hydrogenation of styrene derivatives as studied by ^{1}H-PHIP NMR the shifted resonances are attributed to an initially emerging π-complex between the aromatic portion of the hydrogenated substrate and the catalyst which represents a precursor to the final hydrogenation product prior to its detachment from the catalyst. The kinetic constants for the formation and the decay of these intermediates have been determined for a number of selected substrates using ^{1}H-PHIP NMR spectroscopy [22, 23].

The ^{13}C-PHIP NMR investigations made by Stephan et al. clearly show that it is possible to efficiently transfer polarization from parahydrogen to ^{13}C-nuclei, especially when conducting the experiments at low magnetic fields.

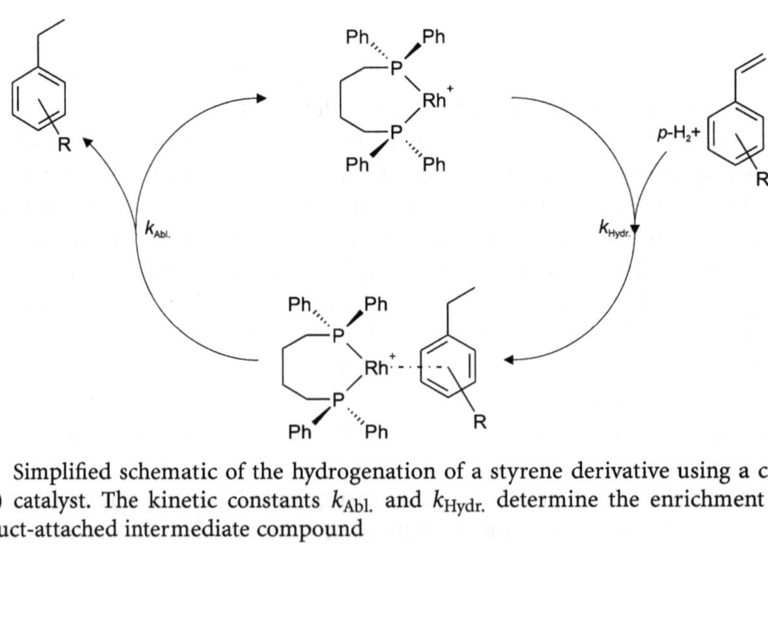

Fig. 5 Simplified schematic of the hydrogenation of a styrene derivative using a cationic Rh(I) catalyst. The kinetic constants $k_{\mathrm{Abl.}}$ and $k_{\mathrm{Hydr.}}$ determine the enrichment of the product-attached intermediate compound

Moreover, it could be shown that the polarization transfer obviously depends on the electronic and steric features of the substituents adjacent to the unsaturated center. The strongest ^{13}C-PHIP enhancements were observed using either 3,3-dimethylbutyne, 1-hexyne, or phenylethyne as the hydrogenation substrate. In these cases, the polarization could be transferred to all carbon nuclei associated with a good to excellent SNR in the respective spectra. A transfer of polarization to all carbon nuclei of the hydrogenation product was also observed when 2-hexyne was used as the substrate, and the same holds true for the substrate molecules 3-hexyne and 4-octyne. From these results it could be derived that the good accessibility of the terminal triple bond of 3,3-dimethylbut-1-yne, phenylethyne, and 1-hexyne implies an easy hydrogenation of these substrates leading to a more efficient transfer of PHIP which subsequently results in a higher SNR as compared to that observed in the ^{13}C-PHIP spectra of the hydrogenation products of 2-hexyne and 3-hexyne, which both carry a more central and hence less accessible triple bond. Finally, the higher degree of electron density located around the triple bond of 3,3-dimethylbutyne and phenylethyne facilitates the coordination of these substrates to the metal center of the catalyst, thereby leading to a slightly enhanced SNR as compared to 1-hexyne.

Furthermore, it is worthwhile pointing out that due to the very low natural abundance of ^{13}C-nuclei, practically all observed product molecules contain only one ^{13}C-nucleus. Accordingly, the polarization signals observed in the ^{13}C-PHIP spectra do not originate from one and the same product molecule but from different singly labeled ones. Since the fraction of the product molecules that contains two or even more ^{13}C-nuclei is practically negligible, the observed ^{13}C-PHIP spectra are the result of a superimposition of the spectra stemming from product molecules which only contain a single ^{13}C-nucleus at the respective position. Hence, a transfer of the polarization from the former parahydrogen nuclei to an individual ^{13}C-nucleus cannot occur via a transfer along the backbone of the molecule but it must be caused by direct or indirect coupling mechanisms between the former parahydrogen nuclei, other indirectly polarized protons, and the corresponding ^{13}C-nucleus (vide infra).

Today, the observation of ^{13}C-PHIP is a common phenomenon observed during the parahydrogenation of a large variety of structurally very different ^{13}C-containing compounds [24], and it can easily be accomplished using a conventional NMR spectrometer and a rather inexpensive system of generating parahydrogen-enriched H_2 gas mixtures. The extent of the transfer efficiency and the likelihood of observing ^{13}C-enhancement on every carbon nucleus of a parahydrogenated molecule, however, crucially depends on the hydrogenation conditions and the efficiency of the hydrogenation reactions. Thus, care has to be taken in the planning of any PHIP transfer experiment by adjusting all important experimental parameters in order to maximize the efficiency of this effect (vide infra).

2.4 PHIP Transfer to ^{15}N

Little experimental data has been published dealing with the transfer of PHIP-derived polarization to ^{15}N. The system studied by Natterer and Bargon [12, 13] is the parahydrogenation of acrylonitrile yielding polarized propionitrile (ethyl cyanide) (Fig. 6). The aim of these studies was to investigate the transfer of the polarization stemming from parahydrogen to ^{15}N-nuclei under various conditions. Figure 7 shows the ^{1}H-PASADENA PHIP spectrum acquired during the parahydrogenation of the starting material and Fig. 6

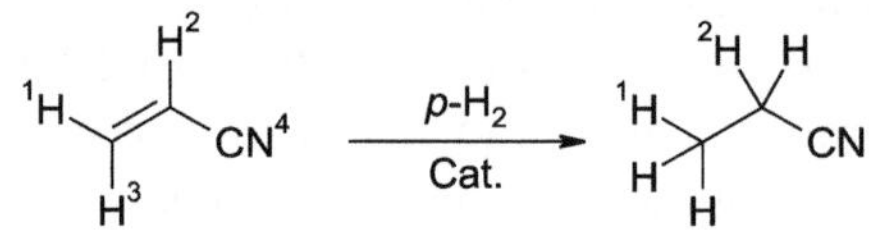

Fig. 6 Reaction scheme of the parahydrogenation of acrylonitrile yielding ^{1}H- and ^{15}N-hyperpolarized propionitrile. Coupling parameters (Hz) for the starting material are $^{3}J_{12} = 11.7$; $^{2}J_{13} = 1.1$; $^{3}J_{23} = 17.7$; $^{3}J_{24} = 1.8$ and for the product $^{3}J_{HN} = 3.2$; $^{3}J_{HH} = 7.8$

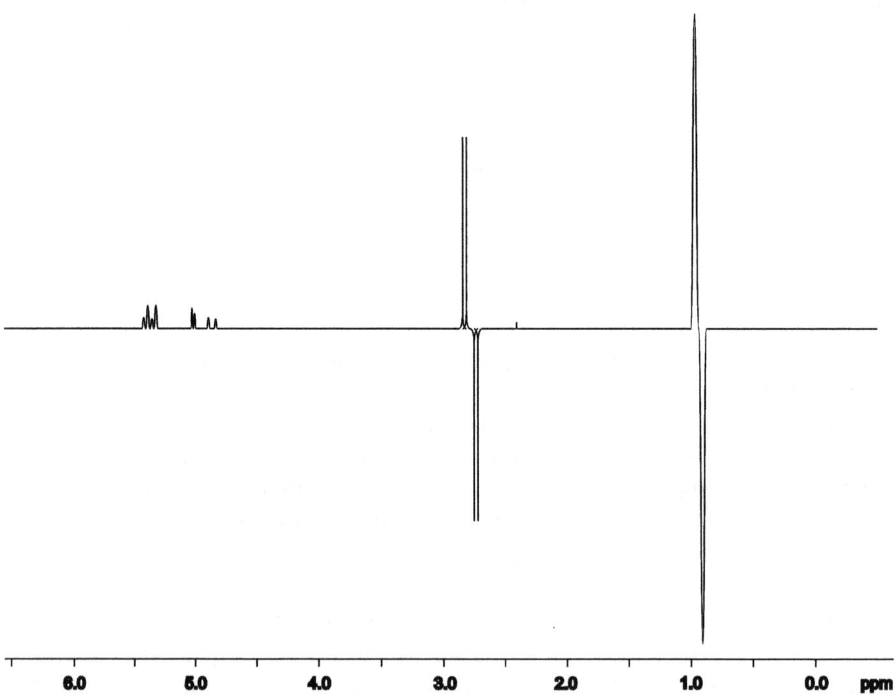

Fig. 7 ^{1}H-PASADENA PHIP spectrum recorded with a single scan 1 s after the hydrogenation reaction of acrylonitrile to propionitrile had been initiated

displays the reaction chemistry of this hydrogenation together with a list of all relevant coupling parameters of the starting material and the product. In Sect. 4.1.1 we will show how specific pulse sequences can be employed to selectively transfer coherence stemming from the parahydrogen molecule to the ^{15}N nucleus present in the parahydrogenation product at natural abundance.

2.5 PHIP Transfer to ^{19}F

Due to its unique physical properties ^{19}F has proven to be a very suitable nucleus for the structural analysis of organic compounds using NMR spectroscopy because its relative sensitivity is almost as high as that of ^{1}H. Furthermore, ^{19}F does normally not occur in organic molecules and this low abundance in naturally occurring substances makes it thus a very specific and also very sensitive probe for the elucidation of molecular structure when specifically labeled samples are used. The increased chemical shift dispersion of ^{19}F compared to ^{1}H and the reduced spectral crowding arising from just one or a few strategically placed fluorine probes in a molecule also offers considerable advantages over more conventional NMR methods.

In a very recent study we demonstrated that the transfer of PHIP to ^{19}F is feasible using a set of singly fluorinated styrene and phenylacetylene derivatives [15]. Moreover, we took a closer look at the mechanisms that are responsible for the transfer of PHIP to heteronuclei using ^{19}F as a very sensitive probe of changes in its nearby electronic environment. We then tried to show whether a quantitative analysis of our results is capable of either confirming or ruling out a dipolar through-space or a scalar through-bond transfer mechanism.

Preliminary experiments were conducted in order to explore whether the transfer of PHIP to fluorine nuclei is observable during the parahydrogenation reactions depicted in Fig. 8. This feasibility study was necessary since there was no evidence so far that parahydrogen-derived polarization transfer to ^{19}F nuclei actually occurs. Our studies, however, clearly confirmed the efficient transfer of PHIP to fluorine nuclei of the selected substrates following their parahydrogenation. In the ^{19}F PHIP spectra of all parahydrogenated substrate molecules the polarization signals representing the single ^{19}F nucleus showed net emission and were characterized by a substantial signal enhancement of about two orders of magnitude compared to a normal ^{19}F NMR spectrum of the same compounds acquired after their hydrogenation using thermally equilibrated dihydrogen gas. Figure 9 summarizes the extent or efficiency of PHIP transfer to the ^{19}F nuclei in terms of the measured enhancement factors (SE) of the obtained polarization signals. As can be seen in this figure, we were not only able to observe PHIP transfer to ^{19}F under low-field conditions using an ALTADENA setup but also under high-field conditions. This observation thus proved that the transfer of PHIP-derived high

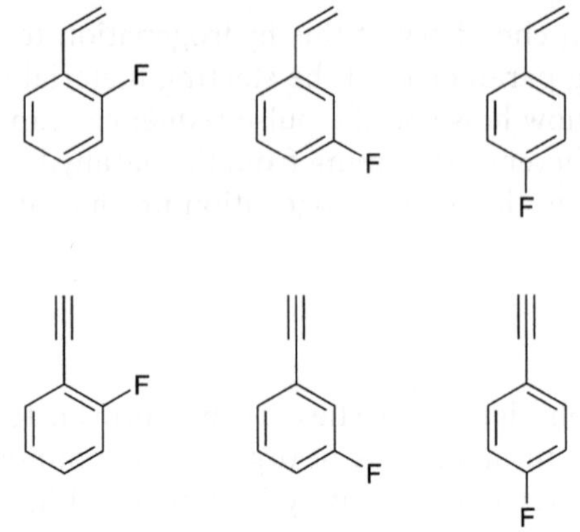

Fig. 8 Structures of the parahydrogenation substrates employed in the ^{1}H–^{19}F polarization study carried out by our own group

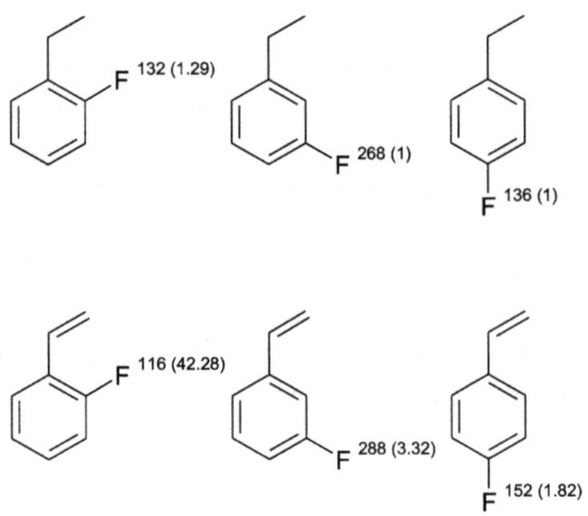

Fig. 9 ^{19}F PHIP signal enhancement factors attained for the singly labeled parahydrogenation products of the fluorostyrene derivatives using ALTADENA (PASADENA) conditions

spin order to a heteronucleus of the parahydrogenation product is also possible at high magnetic fields using PASADENA conditions which yield weakly coupled spin systems with respect to the transferred protons.

Furthermore, a comparison of all experimentally determined signal enhancement factors attained under ALTADENA conditions was carried out and showed that PHIP transfer to ^{19}F is very efficient and leads to a substantial enhancement of more than two orders of magnitude. However, these ALTADENA results did not suggest a direct distance dependency of the transfer efficiency between the former parahydrogen protons and the fluorine nucleus. Rather, our ALTADENA results showed that the highest signal enhancements can be observed for the parahydrogenation product of

3-fluorophenylacetylene and 3-fluorostyrene. In principle, this finding could be a consequence of relaxation processes. A comparison of the measured spin-lattice relaxation times (T_1) of all ^{19}F nuclei in the resulting parahydrogenation products, however, showed that they are virtually identical, and thus relaxation alone cannot account for the different amounts of polarization transfer observed for every individual fluorine nucleus. Nonetheless, this observation could still be well explained and understood in terms of different hydrogenation reaction rates for the different isomers. Applying the concept of a linear free energy relation one can see that the unsaturated bond of the alkene or alkyne portion of these 3-fluoroarene derivatives contains more electron density than that of the 2-fluoro- and 4-fluoroarene derivatives. Therefore, the 3-fluoroarenes will act as much better ligands for the catalyst in the hydrogenation cycle. Accordingly, hydrogenations of these two compounds are postulated to be thermodynamically more favorable than those of the *ortho*- or *para*-substituted isomers, and this will thus lead to an increase of the hydrogenation rate constant. Hence, the amount of polarization transferred to the heteronucleus will also be more profound in the hydrogenation products of the 3-fluoroarenes compared to those of all other isomers and hence should result in a higher SE in the resulting NMR spectrum.

Another result of our study was that the parahydrogen-derived polarization transfer to heteronuclei does not only occur under ALTADENA conditions at low fields but is also present when the whole parahydrogenation reaction is carried out in the strong field of the NMR spectrometer. Furthermore, the observation of PHIP transfer to ^{13}C heteronuclei occurring under PASADENA conditions could also be verified for the fluorostyrene and fluorophenylacetylene derivatives used in this study. A quantitative comparison, however, between the individual ^{19}F polarization signals attained under ALTADENA and PASADENA conditions clearly shows that transfer in the case of our low-field setup is much more efficient. Only for the three phenylacetylene derivatives can a PHIP transfer be observed for all ^{19}F nuclei under low-field conditions. Furthermore, we were able to explain this difference in transfer efficiency between low-field and high-field PHIP using the concept of "isotropic mixing" (vide infra).

2.6 PHIP Transfer to ^{29}Si

The first transfer of PHIP to ^{29}Si was reported by Haake et al. in 1996 during the parahydrogenation of (trimethylsilyl)acetylene yielding ^{29}Si-hyperpolarized (trimethylvinyl)-silane [10] (Fig. 10) using a mononuclear Rh(I) complex as the activating catalyst. This transfer was not just achieved applying a PH-INEPT pulse sequence immediately after the hydrogenation of the starting material under PASADENA conditions (vide infra) but also when the parahydrogenation reaction was carried out in the absence of the

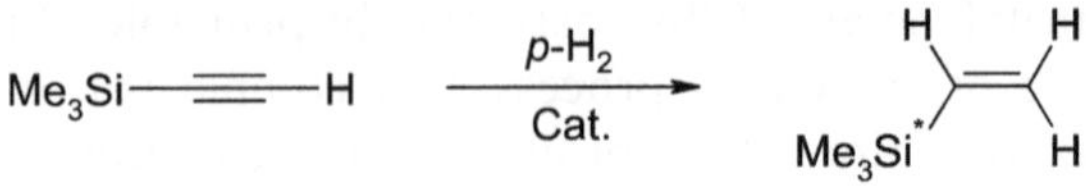

Fig. 10 Parahydrogenation reaction of (trimethylsilyl)acetylene yielding ^{29}Si-polarized (trimethylvinyl)-silane

spectrometer's high field followed by an immediate transfer of the sample to the NMR spectrometer and the subsequent acquisition of the ^{29}Si PHIP spectrum. The attained signal enhancements were substantial and the polarized ^{29}Si signal showed a characteristic antiphase pattern (vide infra) (Fig. 21).

2.7 PHIP Transfer to ^{31}P

The first transfer of PHIP to a heteronucleus was observed in 1989 by Eisenschmid et al. during the oxidative addition of an inorganic iridium(I) complex using parahydrogen gas and applying it in situ to the reaction mixture [7]. In their paper, the authors reported that the parahydrogen-derived high spin order could be selectively and spontaneously transferred to the ^{31}P nuclei of the ligands (dppe, dppv, or dppb). The authors mentioned that the spectra obtained in this fashion were strikingly similar to those using the INEPT pulse sequence even though they suspected that the physical process leading to this polarization transfer is that of the nuclear Overhauser effect (NOE). The reaction chemistry used to develop their "chemical INEPT" experiment corresponds to the oxidative addition of p-H_2 to Ir(I) complexes of the type IrX(CO)(P – P), where "P – P" is either $Ph_2PCH_2CH_2PPh_2$ (dppe), *cis*-$Ph_2PCH=CHPPh_2$ (dppv), or *o*-$(Ph_2P)_2C_6H_4$ (dppb). This reaction proceeds under kinetic control and leads to the stereoselective formation of octahedrally coordinated dihydride complexes as shown in Fig. 11.

When this reaction was carried out in CD_2Cl_2 at 48 °C the transfer of polarization to phosphorus was substantial using ^{31}P NMR spectroscopy, and the two ^{31}P resonances of either the dppe, dppv or dppb ligand showed antiphase polarization with a signal enhancement of ca. 7.4 estimated from

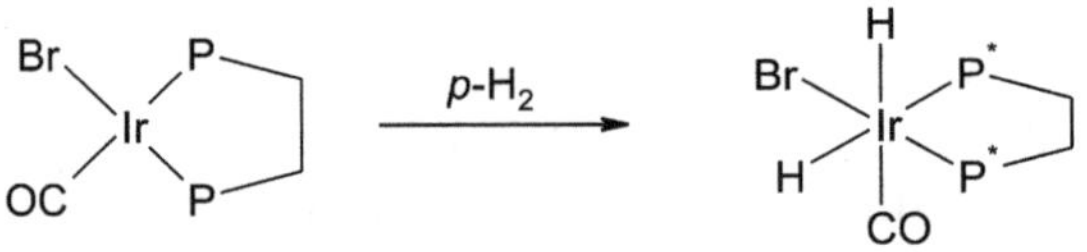

Fig. 11 Oxidative addition of a planar Ir(I) complex using parahydrogen gas mixtures. The reaction yields an octahedrally-coordinated Ir(III) complex whose ^{31}P nuclei benefit substantially from the transfer of PHIP leading to enhanced absorptive and emissive signals in their respective spectra

integration relative to the normal spectrum which was taken several minutes after the reaction had finished.

The results presented by Eisenschmid et al. showed rather impressively for the first time that spin overpopulations generated by the oxidative addition of p-H_2 can be transferred effectively to ^{31}P nuclei resulting in a polarization of the phosphorus nuclei present in one of the ligands of the transition metal complex which subsequently leads to a significant NMR signal enhancement. Additionally, other systems (including Vaska's complex: $IrCl(CO)(PPh_3)_2$) showed similar results regarding polarization transfer and the observation of PHIP transfer could thus be extended to a whole class of transition metal complexes leading to a greater general applicability of the method.

2.8
PHIP Transfer to Other Insensitive Nuclei

Transferring the parahydrogen-derived high spin order of ^1H to heteronuclei is extremely useful when it can be exploited to increase the receptivity of NMR-active nuclei which suffer from an inherently low sensitivity due to a small gyromagnetic ratio and a low natural abundance. In this context, the occurrence of "hetero PHIP" is thus very advantageous and it has been shown that its application can substantially increase the sensitivity of ^2H-, ^{13}C-, ^{15}N-, ^{19}F-, ^{29}Si-, and ^{31}P-nuclei which are present in specifically labeled parahydrogenation precursor molecules [10, 12, 13]. Likewise, the parahydrogen-derived polarization may be transferred to a variety of other NMR-insensitive nuclei provided an unsaturated precursor molecule can be identified that yields the molecule to be polarized upon parahydrogenation.

3
PHIP Transfer Mechanisms

Parahydrogen-Induced Polarization can be transferred to protons and various other types of NMR-active heteronuclei present in the parahydrogenation products. As shown in the previous section this transfer leads to a substantial increase in the signal strengths of these nuclei and facilitates their detection with NMR using a reasonably small number of transients and a concomitant reduction of spectrometer time. Even though the transfer of PHIP to a large number of different heteronuclei could be observed in the past, the mechanisms underlying the process of PHIP transfer are still not fully understood. Furthermore, differing mechanisms have actually been brought forward in the past to explain the phenomenon. In this section we discuss several transfer mechanisms which have been postulated in order to explain the PART phenomenon. Additionally, we present recent experiments carried out by our own group in order to clarify and further elucidate the mechanisms which govern the transfer of PHIP to other protons and heteronuclei of the parahydrogenation product. In particular, we want to demonstrate that it is important to distinguish between high-field and low-field PHIP to sufficiently explain the underlying mechanisms.

3.1 ALTADENA Experiments

In the stray field of an NMR magnet the chemical shift differences between chemically inequivalent protons of a molecule are rather small; a chemical shift difference of 200 Hz at a field strength of 4.7 T corresponds to about 0.2 Hz in the stray field and to even less in the Earth's magnetic field in which ALTADENA experiments are normally initiated. To a similar extent, in a low magnetic field precession frequencies of individual protons and other heteronuclei within a molecule are much more similar than in the high-field case. Hence, for PHIP experiments carried out in the absence of the strong field of an NMR magnet the situation corresponds to what has become known as "isotropic mixing". According to this concept, transfer of PHIP-derived polarization from protons to other protons and heteronuclei is considered to be an exchange of energy between nuclear spins having different spin temperatures. In order to transfer energy between these species efficiently the resonance condition has to be met. In the strong field of an NMR magnet the precession frequencies of protons differ considerably from those of all heteronuclei; therefore, a polarization transfer from the protons to heteronuclei is less likely to occur. If, however, PHIP experiments are conducted in low magnetic fields, e.g., in the stray field of an NMR spectrometer or in the Earth's magnetic field, the precession frequencies of all nuclei in a molecule will be rather similar, rendering an efficient energy transfer between them more likely and consequently more efficient. In the complete absence of any magnetic field—a condition which could not be satisfied yet experimentally—the concept of individually different precession frequencies of heteronuclei does not apply. In such a case the dominating magnetic coupling between magnetically active nuclei would determine the transfer and the eventual distribution of the resulting hyperpolarization exclusively. Achieving this situation would be very desirable for an efficient transfer between the protons stemming from the parahydrogen molecule and individual heteronuclei to occur since the differences between the precession frequencies of protons and ^{13}C nuclei, for example, in the Earth's magnetic field are still about 1.2 kHz and about 15 kHz in the stray field of the NMR magnet. Hence, if a very efficient isotropic mixing between protons and heteronuclei is desired the stray field of the magnet has to be compensated. This compensation can either be achieved using small compensating magnetic fields as realized in "zero-field" NMR [25–27]. Alternatively, the construction of a so-called "zero-field" box built from magnetic shielding material might be beneficial in order to achieve an even more efficient transfer.

Nonetheless, even when using a common ALTADENA setup polarization is usually detectable both on non-parahydrogen protons and on the heteronuclei of the parahydrogenation product. Quite a few mechanisms can conceivably explain this finding:

Firstly, polarization transfer can be produced by the nuclear Overhauser effect (NOE). The magnitude of the NOE, however, is inversely proportional to the strength of the static magnetic field. Thus, at low magnetic fields outside the spectrometer's high-field region strong NOEs can occur and transfer of PHIP-derived polarization both to the protons and the heteronuclei of the parahydrogenation product is feasible. In order to sufficiently explain this mechanism the parahydrogenation reaction of phenylacetylene (phenylethyne) to styrene (phenylethene) can be considered (Fig. 12).

The PHIP spectrum acquired after the parahydrogenation of this compound shows polarization signals not just for the two attached parahydrogen protons but also for the other proton (H_M) of the double bond which was already present in the starting material. Together with the two incorporated protons this nucleus forms an AMX three-spin type of system (Fig. 13) and significantly benefits from the transfer of polarization.

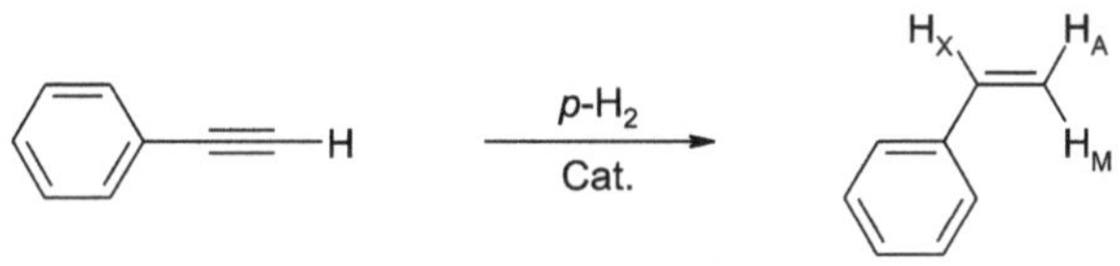

Fig. 12 Schematic of the parahydrogenation of phenylacetylene yielding styrene where the former parahydrogen protons form an AMX spin system together with the proton from the substrate molecule

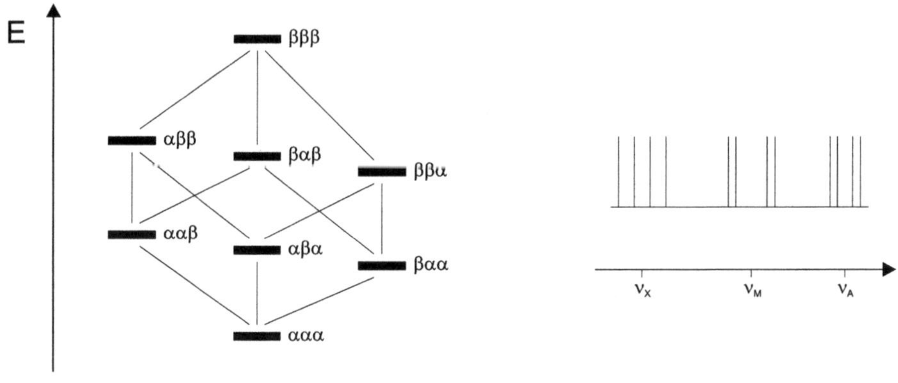

Fig. 13 Schematic representation of the energy levels of a reaction product containing 3 nuclei of different chemical shift of the AMX-type and resulting NMR-spectrum

In order to explain the PHIP spectrum observed during the hydrogenation of phenylacetylene with parahydrogen, cross-relaxation effects have to be incorporated into the theoretical model. These processes are well known in magnetic resonance spectroscopy since they give rise to the NOE. The consequences of various cross-relaxation processes are outlined in Fig. 14: As can be shown by a comparison of the experimental spectrum with simulated PHIP-spectra, the polarization and the PHIP-pattern of the M-resonance in styrene can be explained correctly if a dynamic *scalar* coupling is assumed to induce so-called W_0-transitions between the corresponding energy levels (bottom part). Dynamic *dipolar* coupling gives rise to W_2-transitions which also transfer polarization to the M-group of styrene, but they alone do not interpret the pattern of its PHIP-resonance correctly. This case is illustrated in Fig. 14 in the top part. Only the incorporation of a time-dependent scalar coupling process causing W_0-transitions between energy levels corresponding to a transfer of polarization between H_M and H_X of the product styrene yield an agreement between the simulation and the experiment.

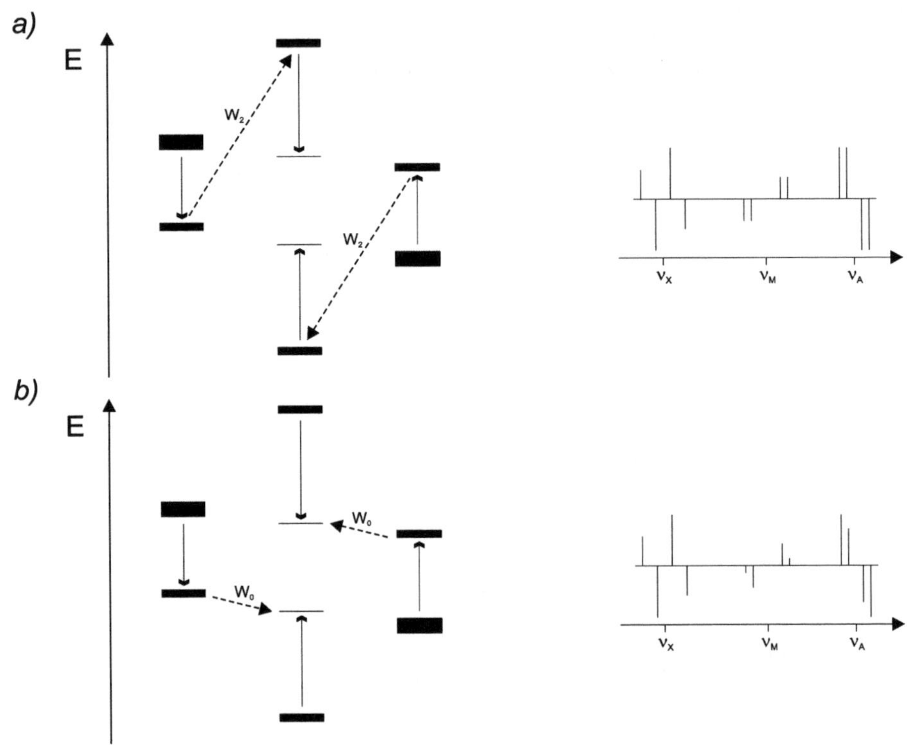

Fig. 14 Representation of the consequences of cross-relaxation processes causing a nuclear Overhauser effect (NOE) in the hydrogenation product styrene during the hydrogenation of phenylacetylene with parahydrogen due to dynamic *dipolar* coupling causing W_2-type transitions (**a**) and subsequent dynamic *scalar* coupling causing W_0-type transitions (**b**)

The origin of this time dependence of the coupling between the vinyl protons in styrene can be explained as follows: The rotation of the phenyl group around the carbon–carbon bond modifies the degree of conjugation between the olefinic and the aromatic double bonds, and, therefore, this motion modulates the scalar coupling constant J between proton H_X stemming from the parahydrogen molecule and proton H_M which originates from the substrate and is, therefore, originally not polarized. This dynamic coupling process transfers spin polarization from the parahydrogen protons to an initially *passive* nucleus of the substrate. Actually, this very type of dynamic coupling can transfer polarization from the protons to any magnetic nucleus to which an appropriate dynamic coupling occurs in the product molecule [28].

3.2 PASADENA Experiments

Whereas both scalar and dipolar coupling mechanisms are responsible for the transfer of PHIP to other protons and heteronuclei of the parahydrogenation product using ALTADENA conditions only the contribution of (dynamic) dipolar coupling seems to play an important part in the case of PHIP transfer occurring under PASADENA conditions. This finding is based upon the results of a recent $^1H-^{19}F$ polarization transfer study carried out by our own group employing a set of singly fluorinated styrene and phenylacetylene derivatives (vide supra) [15]. The fluorinated substrates selected in our study were the three singly fluorinated alkene derivatives 2-fluoro-, 3-fluoro-, and 4-fluorostyrene and the three alkyne derivatives 2-fluoro-, 3-fluoro-, and 4-fluorophenylacetylene. These compounds were chosen since the distance between every single fluorine nucleus and the parahydrogenation center is different in every compound, and the analysis of their polarization spectra can thus serve as a "spectroscopic ruler" when probing different transfer mechanisms. A quantitative analysis of the attained polarization signals in the parahydrogenated phenylacetylene derivatives using highly reproducible PASADENA conditions gives strong evidence for the assumption that PHIP transfer occurring in the spectrometer's high field is mainly or even almost exclusively governed by a dipolar through-space coupling mechanism since its transfer efficiency seems to correlate with the inverse sixth power of the distance between the hydrogenation center and the receiving heteronucleus. In order to demonstrate this behavior the longitudinal single-spin order values F_Z attained for the ^{19}F nuclei of the three newly formed styrene derivatives were plotted against the mean distance r_{HF} between the parahydrogenation center and the heteronucleus. A function proportional to the inverse sixth power of r_{HF} could be fitted through the experimental values using just one parameter. This behavior is expected for a dipolar cross-relaxation interaction being the dominating mechanism responsible for the transfer of initial longitudinal single-spin order of the attached protons (H_Z) to the ^{19}F nucleus

(F_Z) according to Eq. 1.

$$\sigma_{12} = \frac{\gamma_1\gamma_2\hbar}{10}\left(\frac{\mu_0}{4\pi}\right)\frac{\tau_c}{r^6}\left(\frac{6}{1+\left(\omega_1+\omega_2\right)^2\tau_c^2}-\frac{1}{1+\left(\omega_1-\omega_2\right)^2\tau_c^2}\right). \tag{1}$$

In this equation, which represents the general expression of the cross-relaxation rate σ_{12} of a dipolarly coupled spin system comprising two nuclei 1 and 2, τ_c is the rotational correlation time of the individual molecule and ω the angular Larmor frequency of the nuclei observed.

Our results clearly confirmed that dipolar cross-relaxation seems to be the dominating mechanism for the occurrence of PHIP transfer using PASADENA conditions, and this finding thus additionally supports the aforementioned concept of "isotropic mixing".

As a consequence, PHIP transfer occurring under PASADENA conditions is much less efficient compared to ALTADENA PHIP where both mechanisms of dipolar and scalar coupling contribute to the transfer of polarization to heteronuclei. The differences in transfer efficiency are substantial and spectra acquired in both cases can be distinguished easily in terms of different signal-to-noise ratios. As a prime example the ^{13}C-PHIP spectra acquired during the parahydrogenation of 4-fluorophenylacetylene using either ALTADENA (A) or PASADENA (B) conditions are depicted in Fig. 15. Furthermore, as has been found recently, high parahydrogenation rates will significantly in-

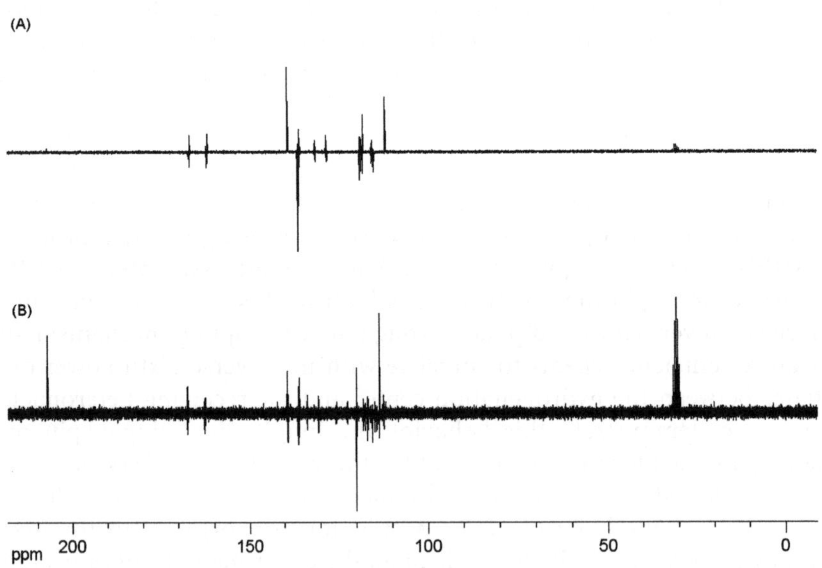

Fig. 15 Comparison of the ^{13}C ALTADENA (**A**) and the ^{13}C PASADENA (**B**) spectrum acquired during the parahydrogenation reaction of 3-fluorophenylacetylene yielding 3-fluorostyrene

crease the likelihood of observing a PHIP transfer to heteronuclei both in the ALTADENA and the PASADENA case.

4 Pulse Sequences

Even though recent results showed that the PART phenomenon can be observed using high-field hydrogenation conditions (PASADENA), a *spontaneous* transfer of parahydrogen-derived polarization does almost exclusively occur when the hydrogenation reaction is initiated in the absence of the high field of the NMR spectrometer (ALTADENA). However, over the years a number of pulse sequences have been devised in order to induce a transfer of PHIP-derived polarization from ^{1}H to heteronuclei in the case that high-field conditions are employed. All these sequences are based on well-established 1D and 2D pulse programs including conventional coherence transfer schemes. In this section we give an outline of the different families of experiments that have been devised in recent years to efficiently induce a PHIP transfer to heteronuclei (X) in the high-field case when weakly coupled spin systems dominate the pattern of ^{1}H–X correlations.

4.1 INEPT

The basic INEPT experiment (Insensitive Nuclei Enhanced by Polarization Transfer) was originally developed to increase the signal strength for nuclides with a low gyromagnetic ratio and a low natural abundance [29, 30]. In this sequence the sensitivity enhancement is usually achieved by transferring magnetization from protons (H) to the individual heteronucleus (X) via spin-spin coupling. Bargon and co-workers have adapted the INEPT sequence to parahydrogen-induced polarization for use when the two protons originating from p-H_2 are in chemically inequivalent positions [10]. Their approach yielded the sequences PH-INEPT, PH-INEPT+, and PH-INEPT($+\pi/4$) (Fig. 16).

In the PH-INEPT and PH-INEPT+ sequences the first proton $\pi/2$ pulse was replaced by a $\pi/4$ pulse with the same phase as the final proton $\pi/2$ pulse compared to a conventional INEPT experiment. The PH-INEPT sequence yields antiphase signals on the heteronuclei when starting with the parahydrogen polarization. Furthermore, it suppresses all signals other than those stemming from parahydrogen-labeled systems. If inphase signals of the heteronuclei are required, the couplings to the heteronuclei can be refocused during one additional delay which can easily be incorporated into the sequence. The PH-INEPT+ scheme contains exactly these additional steps to

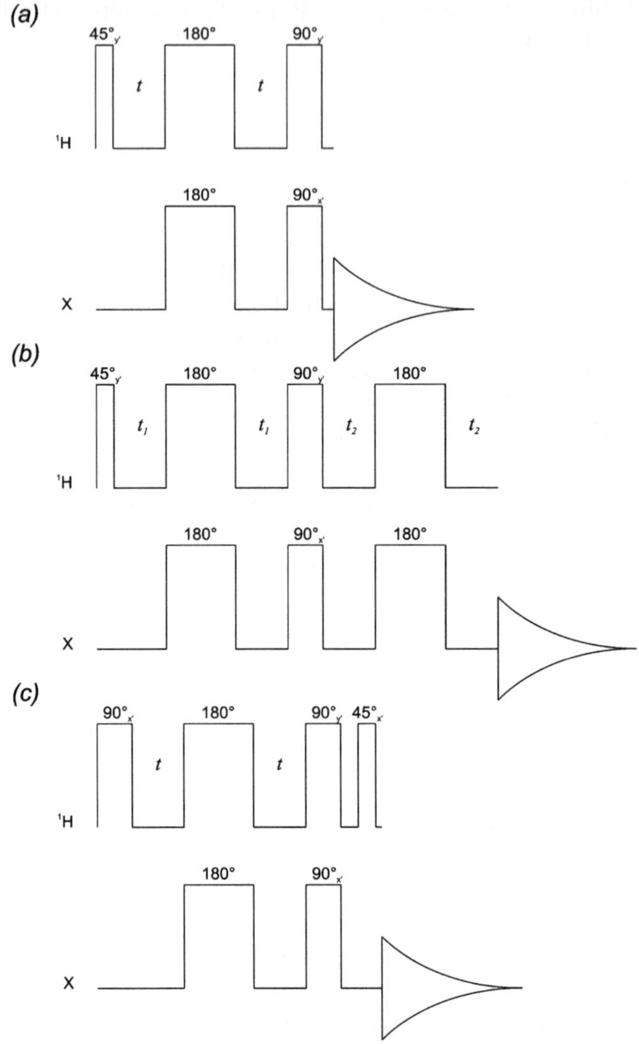

Fig. 16 Graphical representation of the pulse sequences **a** PH-INEPT, **b** PH-INEPT+, and **c** PH-INEPT(+ $\pi/4$)

refocus the antiphase heteronuclear couplings which subsequently enables a decoupling of the protons.

However, the size of the observed heteronuclear signals generated with the PH-INEPT and PH-INEPT+ sequences strongly depends on the ^1H–^1H coupling which is often very small so that optimization is hard to achieve. This characteristic problem is to be expected for some organometallic dihydrides and for the case of non-vicinal hydrogenations. The third INEPT-based sequence designed by Bargon and co-workers called PH-INEPT(+$\pi/4$), however, elegantly overcomes this problem and allows the detection of heteronuclei even when J_{HH} is very small. In this sequence a proton $\pi/4$ pulse is applied at the end of a normal INEPT transfer sequence to convert normally unused trilinear terms into observable magnetization. Eventually, this sequence yields signals that are antiphase with respect to the couplings between protons and heteronuclei and thus no homonuclear proton coupling is required. Furthermore, conveniently short delay times can be used as the observable term evolves with the sine of the sum of all couplings present in the system.

Both the INEPT and INEPT(+ $\pi/4$) sequences were shown to facilitate the detection of ^{2}H, ^{13}C, and ^{29}Si signals upon the Rh(I)-catalyzed hydrogenation of 1,4-diphenylbutadiyne and (trimethylsilyl)acetylene.

4.1.1
^{15}N Spectroscopy with the PH-INEPT Sequence

The system under investigation is the parahydrogenation of acrylonitrile yielding hyperpolarized propionitrile. The general coherence transfer function of the PH-INEPT sequence for a spin system consisting of N nuclei is given by Eq. 2.

$$\sin\left(\pi J_{12}\tau\right)\sum_{i=1}^{2}\left(\prod_{j=4}^{N}\cos\left(\pi J_{ij}\tau\right)\right)\sin\left(\pi J_{iS}\tau\right)I_{iz}S_x\,. \tag{2}$$

In this 1 and 2 symbolize the two former parahydrogen nuclei and the receiving target nucleus, i.e., the nucleus onto which the coherence is transferred, is denoted by S. For the same system at thermal equilibrium, the coherence transfer function for the INEPT sequence is given by Eq. 3.

$$\cos\left(\pi J_{12}\tau\right)\sum_{i=1}^{2}\left(\prod_{j=4}^{N}\cos\left(\pi J_{ij}\tau\right)\right)\sin\left(\pi J_{iS}\tau\right)I_{iz}S_x\,. \tag{3}$$

Quite frequently, the ^{1}H–X coupling (X = heteronucleus) that can be used for the coherence transfer is significantly larger than any coupling between protons. Therefore, a good guess for the optimum delay time τ in the expression in Eq. 2 is obtained by neglecting all cosine terms and setting the delay time to $\tau = 1/\left(2J_{1S}\right)$ or $\tau = 1/\left(2J_{2S}\right)$, depending on which coupling is larger. If polarization transfer is attempted with the PH-INEPT sequence, however, the ^{1}H–^{1}H couplings are more important. Due to the sine dependence of the coherence transfer function on the coupling between protons (Eq. 2), no reasonable delay time is obtained if this homonuclear proton coupling is neglected. Hence, in order to get the optimum delay time one has to plot the actual coherence transfer function and determine its maxima.

In the case outlined here, the ^{1}H–X coupling that is used for the coherence transfer is even smaller than the ^{1}H–^{1}H couplings. Consequently, also for thermal systems in conjunction with the INEPT experiment, a numerical search for the optimum delay time has to be carried out. By examining the plots for the individual coherence transfer functions it turns out that the signals of acrylonitrile and propionitrile have different signs at a specific delay time chosen. This prediction is confirmed by the experiment. The results thus confirm that the outcome of PT experiments can be predicted quantitatively with the methods outlined here.

4.1.2
^{31}P Spectroscopy with the PH-INEPT(+ $\pi/4$) Sequence

The general coherence transfer function of the INEPT(+ $\pi/4$) sequence is given by Eq. 4. All couplings other than those of the two former parahydrogen nuclei to the target heteronucleus have a cosine dependence; they can thus be neglected if J_{1S} or J_{2S} is sufficiently large. It should be emphasized that the coherence transfer function does not depend on the homonuclear proton coupling that exists between the two former parahydrogen nuclei.

$$\prod_{i=1}^{2}\prod_{j=4}^{N} \cos\left(\pi J_{ij}\tau\right) \sin\left(\pi\left(J_{1S}+J_{2S}\right)\tau\right) I_{1z}I_{2z}S_x \, . \tag{4}$$

The system under investigation is the dihydrido compound Vaska's complex, $IrH_2(CO)CI(PPh_3)_2$. The attempt was made to transfer the polarization of the protons that are directly attached to the metal to the ^{31}P nuclei of the ligands. As will be outlined later on in this section, the INEPT(+ $\pi/4$) concept is much better suited for coherence transfer in this system than the PH-INEPT sequence. The signal phase behaves exactly as expected. With prolonged delay time, however, relaxation takes its share. One more feature that is evident is the ability of the parahydrogen labeling method to identify the desired species very quickly.

The experiments with the INEPT(+ $\pi/4$) sequence display only one peak stemming from the dihydrido complex. The INEPT experiment, however, displays three peaks. One of them is easily identified by comparison with a spectrum of the starting materials showing that it belongs to Vaska's complex. The origin of the other two peaks, however, is unknown. The peak intensities in these spectra give no indication of actual concentrations, as all species have different coherence transfer functions.

4.1.3
^{2}H Spectroscopy with the PH-INEPT(+ $\pi/4$) Sequence

One final example is chosen to illustrate the result of an INEPT(+ $\pi/4$) experiment if the couplings of the heteronucleus to the two former parahydrogen nuclei are not equal. The system under investigation is the hydrogenation of

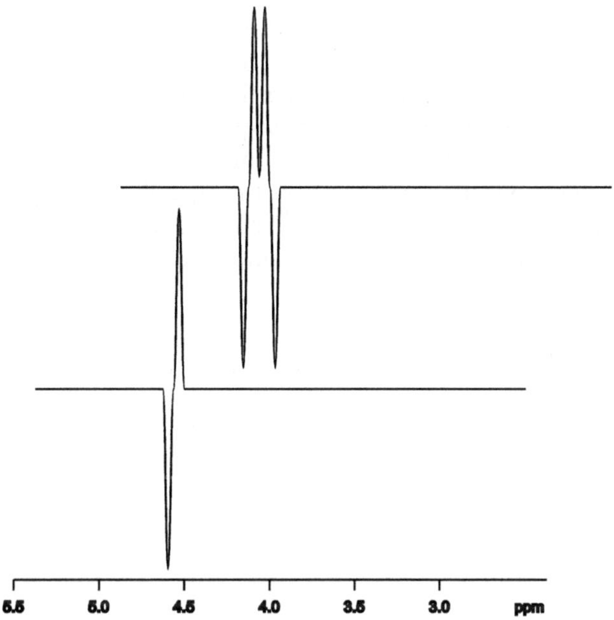

Fig. 17 Simulated ^{2}H single scan spectra recorded during the parahydrogenation of phenylacetylene-d_1 to *cis*-styrene-d_1 using Wilkinson's catalyst, Rh(PPh$_3$)$_3$Cl. *Lower spectrum*: PH-INEPT sequence, *upper spectrum*: PH-INEPT(+ $\pi/4$) sequence

phenylacetylene-d_1 to *cis*-styrene-d_1. The simulated spectra shown in Fig. 17 can be obtained using either the PH-INEPT(+ $\pi/4$) sequence (upper spectrum) or, alternatively, the PH-INEPT sequence (lower spectrum). Concerning the upper spectrum it is worth pointing out that the signal obtained is not a wrongly phased signal but that it does indeed reflect the antiphase multiplet with the theoretically predicted intensities – 1;1;1 – 1. These results could subsequently also be verified experimentally.

4.2 SEPP

The process of devising the PH-INEPT family of pulse sequences, which achieve polarization transfer with the magnetization created by the PASADENA experiment initially, can be quite cumbersome. It was thus very quickly discovered that it would be desirable to find a way of transferring the PASADENA magnetization into the standard magnetization used for any conventional pulse sequence. If this was achieved, a much larger number of NMR pulse sequences could be used in conjunction with parahydrogen. The family of SEPP (Selective Excitation of Polarization using PASADENA) sequences, which was developed at around the same time as the PH-INEPT series, overcomes this problem by removing the need for the application of a $\pi/4$ pulse [11]. This much more versatile method was applied to a number of different standard pulse schemes yielding the pulse sequences SEPP-INEPT, SEPP-INEPT+, and SEPP-DEPT. In these sequences, which are all suitable for inducing a PHIP transfer from protons to heteronuclei of a parahydrogenation product, the application of a selective $(\pi/2)_y$ pulse to the p-H_2 enhanced proton resonance A converts the $2I_z^A I_z^X$ starting state into $2I_x^A I_z^X$ where it evolves under the influence of chemical shift and couplings. If the spin sys-

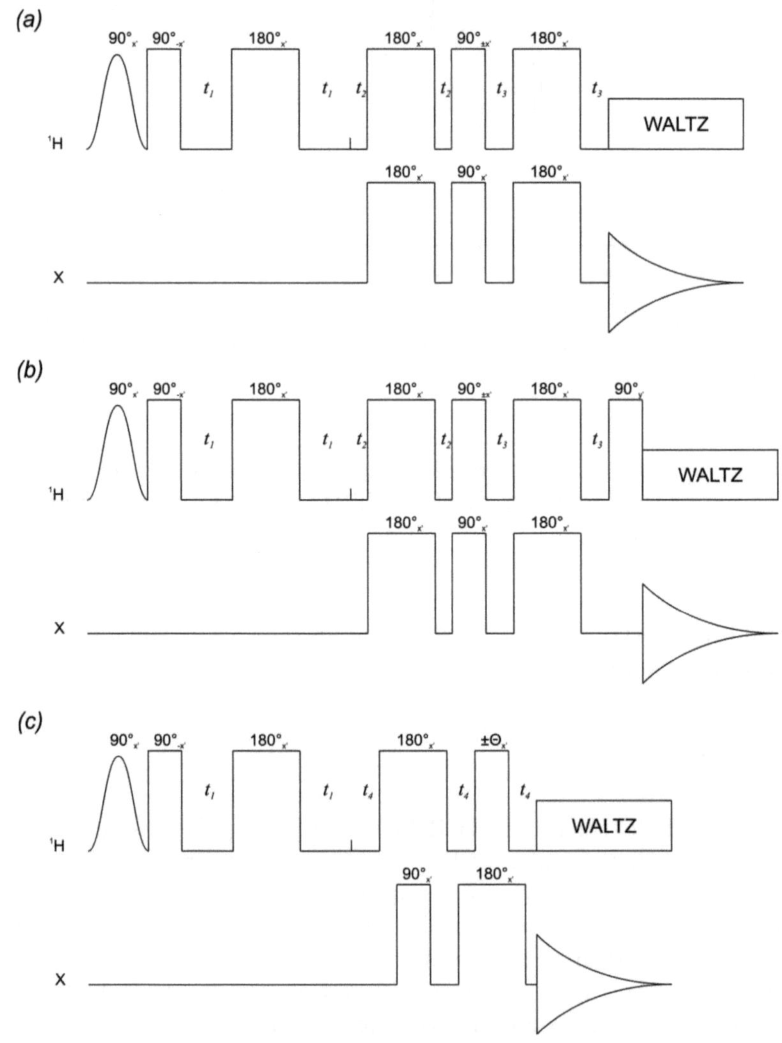

Fig. 18 The three proposed pulse sequences using a selective excitation: **a** SEPP-INEPT, **b** SEPP-INEPT+, and **c** SEPP-DEPT. In each case the delay t_1 is adjusted to $1/(2J_{CH})$

tem represents an AMX group then four time-dependent terms are created by this process. The key I_y^A term corresponds to inphase proton magnetization for resonance A. Optimization of this term yields the largest inphase signal strength for A. The term can subsequently be used in conjunction with SEPP-COSY, SEPP-TOCSY, and SEPP-HOHAHA sequences to probe proton connectivities in hydrogenation products. As these experiments are selective, only p-H_2 enhanced signals are examined and products can be detected in otherwise congested regions of the proton spectrum. The SEPP-TOCSY sequence was tested successfully by acquiring a spectrum for 1-hexene, which contained all the expected proton signals even though only one p-H_2 enhanced resonance was selected. The SEPP-HOHAHA sequence was shown to facilitate the selective probing of resonances within the organic hydrogenation product exploiting a multi-step transfer.

In the case of an AMX spin system, where the X nucleus corresponds to carbon, selective excitation at A converts the $2I_z^AI_z^M$ starting state into $2I_x^AI_z^M$ which is converted into $2I_z^AI_x^M$ by application of a hard $(\pi/2)_y$ pulse. Subsequently, this term evolves under the influence of couplings and chemical shift in the normal way and a $-I_y^M$ term is created after J_{AM} refocusing; the A–X coupling is refocused by a π pulse at the center of the $1/(2J_{AM})$ delay. The SEPP-INEPT, SEPP-INEPT+, and SEPP-DEPT sequences were demonstrated with styrene, the hydrogenation product of phenylacetylene. The new approach generated inphase heteronuclear signals, and the corresponding spectra could be proton decoupled to further increase the sensitivity of the experiment.

4.3 INADEQUATE

After having shown that the acquisition of PHIP in situ spectra of rare and insensitive heteronuclei using just a single scan is a rather straightforward procedure, the question arose whether experiments analogous to the INADEQUATE experiment could also be conducted using parahydrogen. Traditionally, the INADEQUATE (Incredible Natural Abundance Double Quantum Transfer Experiment) pulse sequence was used to evaluate ^{13}C–^{13}C connections directly [31]. As the occurrence of ^{13}C–^{13}C groupings is 100 times lower than the basic ^{13}C level ("isotropic dilution") this type of experiment is extremely insensitive, and it is about a factor 80 000 less sensitive than a ^1H spectrum even if proton excitation is employed.

Natterer et al. have developed a PH-INADEQUATE pulse sequence to achieve the same result with a higher intensity and a minor amount of spectrometer time using parahydrogen-derived ^{13}C polarization [12, 13] (Fig. 19). In this sequence the antiphase magnetization stemming from p-H_2 is selectively and exclusively transferred to pairs of ^{13}C-nuclei where it yields antiphase signals. Thus, in contrast to all other pulse sequences that detect

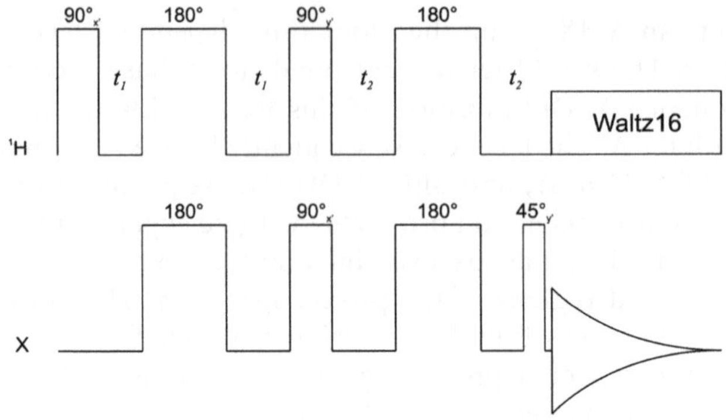

Fig. 19 Pulse sequence for the PH-INADEQUATE experiment

connectivities of nuclei with low natural abundance no unwanted coherence is created in this case and additional suppression techniques are not needed. Additionally, the PH-INADEQUATE approach does not rely on a coupling between individual protons.

When this approach was applied to the examination of the parahydrogenation products of 1,4-diphenylbuta-1-en-3-yne the evaluation of ^{13}C–^{13}C couplings was achieved in 16 scans and one minute of data acquisition. When the INADEQUATE sequence was applied to a standard sample using the same concentration no signal was detected even after an acquisition time of 15 hours. These results thus confirmed the utility of the p-H_2 effect as a means to rapidly examine hydrogenation products also when PASADENA hydrogenation routes are applied.

4.4 Second-Order Effects in Using Pulse Sequences on Reacting Systems

When devising and applying pulse sequences one generally makes two tacit assumptions: (1) the system is strictly weakly coupled and (2) all pulses applied to the system do exactly what they are supposed to do. Indeed, assumption (1) is correct in most cases and will be even more so for NMR investigations carried out at very high magnetic fields. For assumption (2), however, exactly the opposite trend is to be expected and this is what this section is about.

Experimental r.f. pulses have a finite duration. The spectral range covered by an r.f. pulse is inversely proportional to its duration. Consider a proton 90° ($\pi/2$) pulse lasting 5 microseconds, which translates into 20 microseconds for a 360° (2π) pulse. The field strength, $\omega/2\pi$, of this pulse is $1/20 \times 10^{-6} = 50$ kHz. The offset Ω is defined as the difference between the frequency at which the pulse is applied, $\omega_{\mathrm{r.f.}}$, and the resonance frequency, ω, of the nucleus of interest, $\Omega = \omega - \omega_{\mathrm{r.f.}}$.

If a pulse is applied "off-resonance", the axis about which the rotation occurs forms an angle θ with the z-axis with $\tan\theta = \omega/\Omega$. In addition, the flip angle β_{eff} that a spin experiences is given by $\beta_{\text{eff}} = \omega_{\text{eff}}\tau$, where τ is the pulse length and $\omega_{\text{eff}} = \left(\omega^2 + \Omega^2\right)^{1/2}$. As a rule of thumb, the effect of an r.f. pulse can be considered ideal, if $\omega/\Omega > 10$. For short proton pulses, this condition easily holds: The maximum offset for a standard proton spectrum is about 10 ppm, corresponding to 2 kHz on a 200 MHz spectrometer. Frequently, however, one deals with longer pulses like, for example, in TOCSY experiments to avoid sample heating or on older instruments if the proton pulses are applied through the proton decoupler during heteronuclear polarization transfer experiments. For a $\pi/2$ pulse of 30 microseconds the field strength reduces to about 8.3 kHz. Consequently, if a pulse of such length is applied some ppm "off resonance", the effects described become significant. These considerations are displayed graphically in Fig. 20.

In the pulse sequences shown here, it is usually expected that the INEPT sequence that is used in these experiments yields antiphase signals for systems at thermal equilibrium, whereas no observable signals are formed for parahydrogen-labeled systems. If, however, pulses are applied considerably "off-resonance" experimental artifacts usually occur. A striking feature which occurs in these spectra very often is the large signal of the product that is obtained even though the "wrong" pulse sequence is applied, i.e., a pulse sequence that should not affect any polarization transfer to heteronuclei in

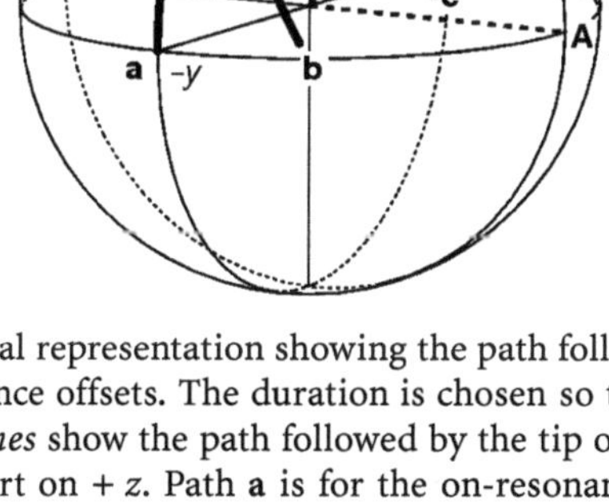

Fig. 20 Three-dimensional representation showing the path followed during an x pulse for various different resonance offsets. The duration is chosen so that, on resonance, the flip angle is 90°; the *thick lines* show the path followed by the tip of the magnetization vector which is assumed to start on + z. Path **a** is for the on-resonance case; the effective field lies along x and is indicated by the *dashed line* **A**. Path **b** is for the case where the offset is half the RF field strength ($\Omega = \omega_1/2$); the effective field is marked **B**. Paths **c** and **d** are for offsets equal to 1.5 times the RF field strength, respectively; the effective field directions are labeled **C** and **D**

parahydrogen-labeled compounds. This effect can be rationalized by inspection of Fig. 20. Trajectory "b" in this figure represents the situation of a $\pi/2$ pulse with phase x that is applied considerably "off-resonance". The outcome of such a pulse is similar to that of a $\pi/2$ pulse with phase $\varphi = \pi/4$ applied "on-resonance".

Shifting the phase of the last proton pulse of the INEPT sequence by $\pi/4$ effectively transforms the INEPT sequence into the INEPT(+ $\pi/4$) sequence. Consequently, if the pulses are applied "off resonance", instead of applying an INEPT sequence, which does not transfer the parahydrogen polarization to heteronuclei, one actually performs an INEPT(+ $\pi/4$) experiment, which enables polarization transfer. If the pulses are applied "on-resonance" the behavior of the spin systems shifts towards what is expected for ideal pulses, i.e., the product signals get smaller, whereas the signals of the starting material increase.

Still, even if the pulses are applied exactly "on-resonance", sometimes product signals remain detectable. The most likely cause for this occurrence is a polarization transfer via the NOE. When using longer delay times, even greater product signals are obtainable. Prolonging the delay time by a factor of 5 to 200 ms has a dramatic effect on the polarization signals, while the signals of the starting material remain more or less unaltered.

Using even longer delay times (400 ms) still yields strong polarization signals but also has a distorting effect on the signal shape. Normally, the spectral appearance then resembles remarkably that which is obtained when this reaction is carried out using ALTADENA conditions and applying a single $\pi/2$ pulse on the heteronucleus for detection. This further corroborates the claim that in both cases cross-relaxation effects are responsible for the occurrence of a signal.

When the first pioneers attempted to use pulse sequences in conjunction with parahydrogen they actually employed the "wrong" pulse sequences

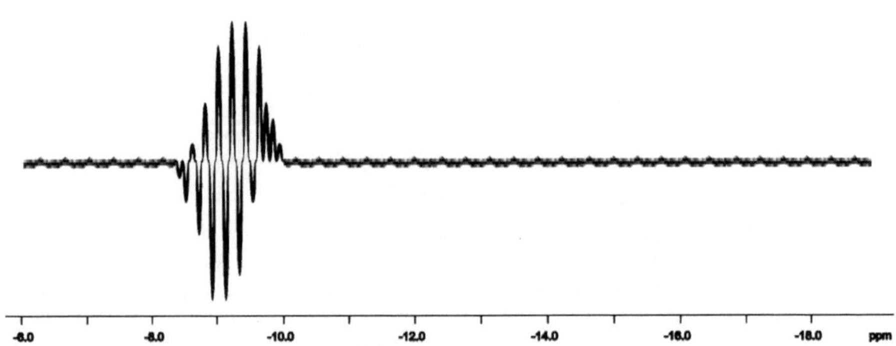

Fig. 21 ^{29}Si NMR spectrum recorded during the hydrogenation of (trimethylsilyl)acetylene to yield (trimethylvinyl)-silane using ALTADENA conditions and acetone-d_6 as the solvent

INEPT and INEPT+. The polarization signals they obtained were strongly distorted and, in addition, the enhancements were always much lower than what was hoped for. In the light of the ideas outlined in this section, such results should no longer be puzzling.

5
Applications of PHIP Transfer

Parahydrogen-Induced Polarization can be transferred to insensitive NMR-active nuclei when certain experimental requirements are met. For example, recent ^{13}C- and ^{19}F-PHIP studies using a specific set of polarizable hydrogenation substrates have shown that it is extremely beneficial to initiate the parahydrogenation reaction at low magnetic fields prior to the acquisition of the NMR spectrum. Also, a large parahydrogenation rate constant $k_{hyd.}$ will eventually lead to the observation of substantial polarization on the heteronucleus of the parahydrogenation product. When symmetric substrates are employed the sensitivity enhancement of the heteronuclei can even further be increased leading to signal enhancements of more than three orders of magnitude as observed in the case of hyperpolarized ^{13}C-nuclei of the parahydrogenation product maleic acid dimethyl ester. Due to the inherent increase in signal intensity associated with the PHIP method the phenomenon has recently found its application in the field of magnetic resonance imaging (MRI) and magnetic resonance tomography (MRT) for the non-invasive and radiationless visualization of tissue and blood circulation in living organisms [32].

The first attempt to exploit PHIP-derived sensitivity enhancement of ^{13}C-nuclei for magnetic resonance imaging techniques was carried out by Golman et al. in 2001 [33]. The investigators employed the parahydrogenation of acetylenedicarboxylic dimethyl ester yielding maleic acid dimethyl ester to generate in vivo ^{13}C angiograms of an anesthetized rat using a single shot RARE sequence on a sub-second timescale. In their experiment a time-dependent diabatic variation of the field strength (field cycling scheme) was applied in order to maximize the PHIP transfer efficiency from hyperpolarized ^{1}H nuclei to ^{13}C. The contrast medium (maleic acid dimethyl ester) used in this experiment fulfilled a number of criteria which proved to be crucial for the generation of an "active" contrast agent suitable for MR imaging using the PHIP transfer phenomenon since it combined the long longitudinal relaxation times (T_1) of ^{13}C with a high concentration of the NMR-active nucleus and a high degree of polarization. Moreover, the long transverse relaxation times (T_2) of the ^{13}C-nuclei present in the product molecule added the opportunity of using multiecho pulse sequences [34, 35] which refocus the transverse magnetization component after each spin echo, thereby utilizing the available signal amplitude more efficiently. The high efficiency of the polarization experiment was due to the application of the aforementioned field cycling scheme whose effects on the spin system of a parahydrogenated molecule can be explained as follows:

The non-equilibrium spin order obtained during the parahydrogenation of a suitable precursor molecule has to be converted into longitudinal polarization before the parahydrogenated compound can be used for an imaging experiment. In the eigenbase of the Hamiltonian of a three-spin system four elements of the spin density matrix are different from zero and only two of them are independent resulting in a zero carbon polarization. Devising a diabiatic field cycling scheme can produce substantial polarization which is observable on the ^{13}C-nucleus. By reducing the external magnetic field strength, the combined proton-carbon spin system is brought into the strong coupling regime, thereby enabling intimate coupling between carbon nuclei and the two protons originating from the parahydrogen molecule. If the magnetic field strength is smaller than 10^{-8} T the eigenstates depend only on the parts of the Hamiltonian that include the scalar couplings and are practically indistinguishable from the zero field states. Above 10^{-5} T the Hamiltonian is dominated by the Zeeman interactions, implying constant eigenstates. There is thus a transition region where the eigenstates continually change from low to high field. Changing the field symmetrically when traversing this region will not accomplish any change to the spin system. If one starts at a high magnetic field and suddenly decreases the field, the spin system will not preserve the original populations of the eigenstates, leading to a nonadiabatic or diabatic process. When the field is now raised adiabatically, the populations of the instantaneous eigenstates remain constant. This field cycling results in a rearrangement of the populations of the original eigenstates so that the system now displays an NMR spectrum where the allowed transitions are predominantly in phase, corresponding to a substantial polarization. The result of the field cycling will in a complicated way depend on the scalar couplings of the system, as well as the precise profile of the cycling of the magnetic field. To optimize the carbon polarization, the critical field and timing will have to be calculated for each case. In the presence of the Earth's magnetic field the spin system evolves under a high field regime, and an instantaneous decrease of the field to below 10^{-7} T is required to obtain the desired redistribution in the populations of the eigenstates. Simulation of the field cycling procedure showed that a polarization of about 50% can be obtained for the system acetylenedicarboxylic dimethyl ester/maleic acid dimethyl ester, neglecting the contribution from relaxation and assuming 100% parahydrogen. A thorough theoretical treatment and explanation of the field cycling phenomena observed when applying PHIP experiments has been developed and is given in various reviews [36–38].

The actual experimental setup for the field cycling apparatus used for the generation of parahydrogen-polarized ^{13}C-containing compounds in combination with MR angiography, which has recently found its first completely automatized application for three-dimensional imaging of hyperpolarized ^{13}C in vivo [39], was first described by Jóhannesson et al. employing the parahydrogenation of hydroxyethyl acrylate yielding hydroxyethyl proprionate [36]:

In their experiment the parahydrogen was produced by passing thermally equilibrated hydrogen gas through a commercially available Rh(I)-catalyst at a temperature of 14 K. The parahydrogen content obtained was found to be larger than 95% by NMR. The apparatus for the field cycling experiments is outlined in Fig. 22.

The process is under computer control and all liquid and gas flows are controlled by solenoid valves. The process works as follows: A reactor chamber is filled with parahydrogen to a pressure of 10 bar. The temperature of the reactor is kept at 334 K, which represents an empirically optimized temperature with respect to the final ^{13}C polarization. A mixture of the substrate and the catalyst is dispensed into the reactor as a narrow jet. After three seconds, the liquid is transferred into the low field chamber with the field initially set to 10^{-4} T. After a delay of 0.5 s the field is taken down to about 30 nT in 1 ms. The field is then ramped up exponentially to 10^{-4} T according to the desired time constant, whereupon the sample is ejected and collected in a syringe for a subsequent injection into the object to be scanned.

The low field chamber is a small glass chamber sitting inside a set of two concentric coils, in turn sitting inside a set of three concentric layers of metal to shield this region from surrounding magnetic fields. Careful demagnetiza-

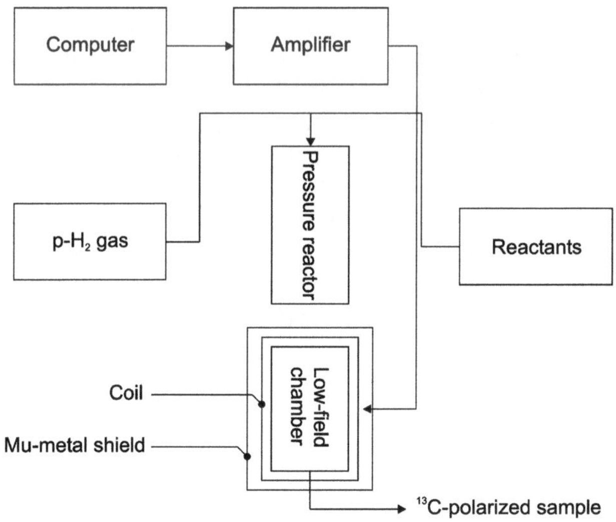

Fig. 22 Schematic representation of the field cycling experiment employed by Jóhannesson et al. for the generation of a parahydrogenated ^{13}C-containing contrast agent used in MR angiography

tion of the metal in a 50 Hz AC field routinely gives a residual field of 10 to 30 nT inside this chamber. The field is controlled by a command voltage fed to a voltage controlled current amplifier. The set of coils are two concentric coils with the current running in opposite directions and balanced to give as close to zero field as possible on the outside. This is necessary to allow for the rapid change of the field inside the conductive metal shields. The chemical reaction employed in this study is outlined in Fig. 23.

An aqueous solution of hydroxyethyl acrylate is parahydrogenated in the presence of a soluble Rh(I) catalyst. The hydrogen is transferred jointly, i.e., without scrambling between molecules, to the substrate to yield hydroxyethyl propionate. For spectroscopic experiments deuterium oxide was preferred whereas for imaging experiments ordinary water was used as the solvent. Normally the experiments were run on a 5 ml scale involving 100 mg of hydroxyethyl acrylate. Immediately after the field cycling, the reaction solution is transferred to either an NMR tube (spectroscopic experiments) or a syringe (imaging experiments). The NMR spectrometer used for the spectroscopic experiments is equipped with an external trigger and set up to start the acquisition without any delay. The time elapsed from the field cycling procedure to the sampling of the spectrum is thereby kept to a minimum of about 7 s. The free induction decay (FID) in the spectroscopic experiment is acquired after applying a 30° r.f. read pulse. After determining the chemical yield, the ^{13}C-spectrum of the polarized product is referenced to a thermal spectrum of a ^{13}C-labeled sample of known concentration, collected with the same parameter settings. This makes it possible to calculate the signal enhancement and the polarization of the reaction product.

Using a remagnetization time of 1.2 s an optimal ^{13}C polarization for hydroxyethyl propionate was obtained. Longer remagnetization times caused a decrease in polarization which was attributed to relaxation of the spin order of the two protons originating from the parahydrogen molecule. The SE of the ^{13}C signal was – 37 900 compared with a thermal equilibrium signal at 7 T and 334 K. This corresponds to a polarization equal to – 21%. The discrepancy between the experimentally obtained polarization and the theoretically predicted polarization corrected for by ^{13}C relaxation (– 25%) is most probably due to proton relaxation during field cycling. The longitudinal relaxation time for the ^{13}C-nucleus of the carbonyl group was found to be 60 s at 7 T

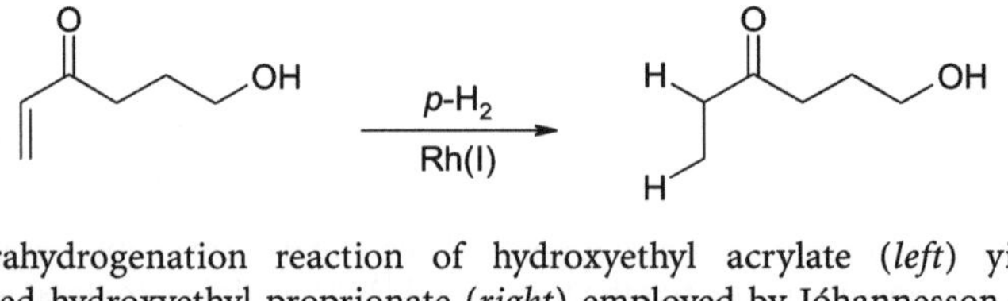

Fig. 23 Parahydrogenation reaction of hydroxyethyl acrylate (*left*) yielding ^{13}C-hyperpolarized hydroxyethyl proprionate (*right*) employed by Jóhannesson et al. during their field cycling experiment

and 334 K. For the in vivo experiment ^{13}C-labeled material was used. The pulse sequence used for data collection was the true FISP sequence which has proven to be suitable for the use in MR angiography of hyperpolarized ^{13}C-containing contrast agents [40].

In an attempt to further increase the applicability of parahydrogenation precursor molecules suitable as active MRI contrast agents, Aime et al. have recently synthesized a new class of long-chained alkyne derivatives attached to a polymeric lysine resin [41]. The compounds depicted in Fig. 24 represent long aliphatic chains with a single alkynic hydrogenation site. Due to the fact that these chains represent large-sized systems in which the hyperpolarizable moieties are linked to the macromolecular backbone by flexible chains that allow ample opportunity for local motions, the rotational correlation time (τ_c) can be kept rather short resulting in long longitudinal relaxation times (T_1) for ^{1}H- and ^{13}C-nuclei. This rather crucial prerequisite of a suitable MRI contrast agent enables these compounds to act as highly potential substrate molecules in future experiments in this field of research.

Moreover, the recently demonstrated feasibility of PHIP transfer to ^{19}F nuclei has extended the number of heteronuclei further whose sensitivity experiences a significant increase upon the addition of parahydrogen to specifically labeled unsaturated substrates. The amount of polarization transferred to ^{19}F using ALTADENA conditions was found to be substantial and the associated signal enhancement factors exceeded two orders of magnitude consistently. This increase in ^{19}F signal intensity opens up a new range of potential applications for hyperpolarized ^{19}F nuclei present in biologi-

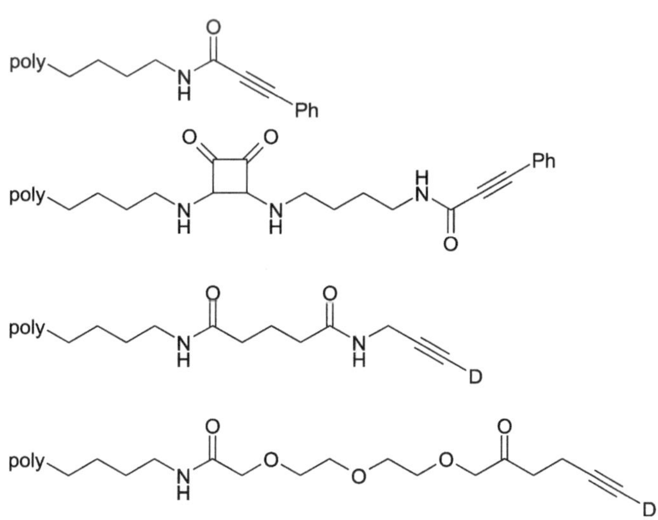

Fig. 24 Structure of the long and flexible parahydrogenation substrates designed by Aime et al. attached to a polymeric lysine resin

cally active molecules. Since the development of new drugs very often involves the use of fluorine as a component to modulate biological activity or other physical properties of these compounds it might be beneficial to exploit this PHIP-derived signal increase for medical imaging techniques as demonstrated by Golman et al. using hyperpolarized ^{13}C nuclei [32]. Specifically chosen molecules might then serve as "active" contrast agents for MRI and MRT experiments following the parahydrogenation of suitable precursor molecules which contain at least one ^{19}F nucleus. Other applications might also involve the use of ^{19}F-hyperpolarized aliphatic blood surrogates in contrast-enhanced magnetic resonance angiography (CE-MRA) as previously shown by Svensson et al. for ^{13}C-hyperpolarized compounds using Dynamic Nuclear Polarization (DNP) [40], and the assessment of tissue blood flow using hyperpolarized ^{19}F-containing tracers [42] might also represent another potential application of this method. However, carefully optimized protocols for the processes yielding ^{19}F-hyperpolarized molecules that are suitable for the injection into living tissue both in terms of biological compatibility and high degrees of hyperpolarization will be necessary to standardize such a procedure.

6 Concluding Remarks

PHIP-NMR spectroscopy of heteronuclei represents an active field of research both in chemistry and medicinal imaging. In this review we have tried to outline the historical development of the technique and its importance for sensitivity enhancement during the observation of insensitive heteronuclei using NMR spectroscopy. Furthermore, we have discussed the dominating mechanisms which are important for the transfer of PHIP to other protons and heteronuclei of a parahydrogenation product, and we have shown that these mechanisms differ in detail depending on whether high-field (PASADENA) or low-field (ALTADENA) hydrogenation routes are applied. The review was concluded by a description of applications of the PART phenomenon in magnetic resonance imaging (MRI) and magnetic resonance tomography (MRT). We believe that the "hetero PHIP" phenomenon will find an even greater variety of important applications in clinical research in the future. These will surely further promote the general applicability of the method.

Acknowledgements We are indebted to Dr J. Keeler (Cambridge University, Cambridge, UK) for allowing us to reproduce Fig. 20. L.T.K. thanks the German National Academic Foundation (*Studienstiftung des deutschen Volkes*) for constant support and generous funding throughout his studies. Further financial support from the Deutsche Forschungsgemeinschaft (DFG) and the Theodor-Laymann-Stiftung (L.T.K.) is also gratefully acknowledged.

References

1. Bowers CR, Weitekamp DP (1986) Phys Rev Lett 57:2645
2. Bowers CR, Weitekamp DP (1987) J Am Chem Soc 109:5541
3. Eisenschmid TC, Kirss RU, Deutsch PP, Hommeltoft SI, Eisenberg R, Bargon J, Lawler RG, Balch AL (1987) J Am Chem Soc 109:8089
4. Natterer J, Bargon J (1997) Prog NMR Spectrosc 31:293
5. Duckett SB, Sleigh CJ (1999) Prog NMR Spectrosc 34:71
6. Pravica MG, Weitekamp DP (1988) Chem Phys Lett 145:255
7. Eisenschmid TC, McDonald J, Eisenberg R, Lawler RG (1989) J Am Chem Soc 111:7267
8. Duckett SB, Newell CL, Eisenberg R (1993) J Am Chem Soc 115:1156
9. Barkemeyer J, Haake M, Bargon J (1995) J Am Chem Soc 117:2927
10. Haake M, Natterer J, Bargon J (1996) J Am Chem Soc 118:8688
11. Barkemeyer J, Bargon J, Sengstschmid H, Freeman R (1996) J Magn Reson A 120:129
12. Natterer J, Barkemeyer J, Bargon J (1996) J Magn Reson A 123:253
13. Natter J (1997) PhD thesis, University of Bonn
14. Stephan M, Kohlmann O, Niessen HG, Eichhorn A, Bargon J (2002) Magn Reson Chem 40:157
15. Kuhn LT, Bommerich U, Bargon J (2006) J Phys Chem A 110:3521
16. Bargon J, Kandels J, Woelk K (1993) Z Phys Chem 180:65
17. Aime S, Canet D, Dastrù W, Gobetto R, Reineri F, Viale A (2001) J Phys Chem A 105:6305
18. Hübler P, Bargon J (2000) Angew Chem Int Ed 39:3701
19. Aime S, Gobetto R, Reineri F, Canet D (2003) J Chem Phys 119:8890
20. Sørensen OW, Ernst RR (1983) J Magn Reson 51:477
21. Haake M, Barkemeyer J, Bargon J (1995) J Phys Chem 99:17539
22. Giernoth R, Hübler P, Bargon J (1998) Angew Chem Int Ed Engl 37:2473
23. Hübler P, Giernoth R, Kümmerle G, Bargon J (1999) J Am Chem Soc 121:5311
24. Bommerich U (2005) PhD thesis, University of Bonn
25. Weitekamp DP, Bielecki A, Zax D, Zilm K, Pines A (1983) Phys Rev Lett 50:1807
26. Zax DB, Bielecki A, Zilm KW, Pines A, Weitekamp DP (1985) J Chem Phys 83:4877
27. Thayer AM, Pines A (1987) Acc Chem Res 20:47
28. Natterer J, Schedletzky O, Barkemeyer J, Bargon J, Glaser SJ (1998) J Magn Reson 133:92
29. Morris GA, Freeman R (1979) J Am Chem Soc 101:760
30. Burum DP, Ernst RR (1980) J Magn Reson 39:163
31. Bax A, Freeman R, Kempsell SP (1980) J Am Chem Soc 102:4849
32. Golman K, Olsson LE, Axelsson O, Månsson S, Karlsson M, Petersson JS (2003) Br J Radiol 76:S118
33. Golman K, Axelsson O, Jóhancsson H, Månsson S, Olofsson C, Petersson JS (2001) Magn Reson Med 46:1
34. Mansfield P, Maudsley AA (1977) J Magn Reson 27:101
35. Hennig J, Nauerth A, Friedburg H (1986) Magn Reson Med 3:823
36. Jóhannesson H, Axlesson O, Karlsson M (2004) C R Physique 5:315
37. Goldman M, Jóhannesson H, Axelsson O, Karlsson M (2005) Magn Reson Imag 23:153
38. Goldman M, Jóhannesson H (2005) CR Physique 6:575
39. Bhattacharya P, Harris K, Lin AP, Mansson M, Norton VA, Perman WH, Weitekamp DP, Ross B (2005) MAGMA 18:245

40. Svensson J, Månsson S, Johansson E, Petersson JS, Olsson LE (2003) Magn Reson Med 50:256
41. Aime S, Dastrù W, Gobetto R, Viale A (2005) Org Biomol Chem 3:3984
42. Johansson E, Olsson LE, Månsson S, Petersson JS, Golman K, Ståhlberg F, Wirestam R (2004) Magn Reson Med 52:1043

Top Curr Chem (2007) 276: 69–110
DOI 10.1007/b135837

Published online: 10 September 2005

Investigations in Supercritical Fluids

Heiko G. Niessen[1] · Klaus Woelk[2] (✉)

[1]Department of Neurology II, Otto von Guericke University, Leipziger Str. 44, 39120 Magdeburg, Germany
heiko.niessen@medizin.uni-magdeburg.de

[2]Department of Chemistry, University of Missouri-Rolla, 1870 Miner Circle, Rolla, MO 65409-0010, USA
woelkk@umr.edu

Abstract In recent years, supercritical fluids (SCF) have drawn substantial interest as modern solvents for chemical reactions, separations, and extractions in the area of basic research as well as in industrial processes. Especially, supercritical carbon dioxide ($scCO_2$) is a desirable replacement for organic solvents because it is inexpensive, nontoxic, nonflammable, environmentally benign, and exhibits ease of recycling and disposal. This contribution provides for an overview of the basic physical properties of SCFs, such as density tuning, diffusivity, viscosity and interaction with fluorinated compounds. In addition, in situ NMR spectroscopy in toroid cavity autoclaves and the specialized technique of parahydrogen-induced polarization are introduced as superior and beneficial for investigating physical properties and chemical reaction in SCFs. The concept of turnover frequencies and active sites in catalysis is reviewed, while specific examples are given for homogeneous hydrogenations and hydroformylations in $scCO_2$. The emerging field of colloid catalysis in SCFs is illustrated with the ultrafast single-phase hydrogenation of alkynes using bimetallic colloids in inverted polymer micelles as the catalyst.

Keywords Supercritical carbon dioxide · Homogeneous catalysis · In situ NMR spectroscopy · Toroid cavity · Colloids

Abbreviations

B_0	Static magnetic field in NMR spectroscopy
B_1	Alternating magnetic field of NMR radiofrequency pulses
BARF	Tetrakis(3,5-bis(trifluoromethyl)phenyl) borate
C_1	Compound with a single carbon atom
D	Diffusion coefficient
E	Energy
k	Boltzmann's constant
MAGROFI	Magnetization-grating rotating-frame imaging
M_n	Average molecular mass in atomic mass units
PEEK	(Polyether)ether ketone
PHIP	Parahydrogen-induced polarization
p_c	Critical pressure
$p(CO_2)$	CO_2 pressure
$p(H_2)$	H_2 pressure
p-H_2	Parahydrogen
r_A	Hydrodynamic radius
scCO_2	Supercritical carbon dioxide
SCF	Supercritical fluid
SCR	Substrate-to-catalyst ratio
t	Time
T	Temperature
T_1	Spin-lattice relaxation time constant
T_c	Critical temperature
TCA	Toroid cavity autoclave
TCD	Toroid cavity detector
TOF	Turnover frequency
TON	Turnover number
η	Dynamic viscosity
ν	NMR resonance frequency

1 Introduction

Technical achievements of the past decades, which paved the way to increased health and prosperity in industrialized countries, have also lead to increased ecological challenges and have called for more sustainable production procedures. Two of the major challenges of modern ecology are the reduction of waste production in industrial processes and the recycling or disposal of existing waste. Especially, the immense use of organic solvents and halogenated chemical reagents as well as the consumption of valuable drinking water in industrial processes have triggered a range of approaches towards more energy efficient and sustainable procedures [1–4]. In this context, supercritical carbon dioxide ($scCO_2$) is often valued as a modern solvent for it is environmentally benign, nontoxic, readily available, inexpensive, and more or less chemically inert. The diffusivities of chemicals dissolved in, or mixed with, $scCO_2$ are usually higher and the viscosities are lower compared with those of conventional organic solvents. Prominently, gases are completely and quickly miscible with $scCO_2$ at almost all ratios. Furthermore, because the dissolution properties in $scCO_2$ vary with the supercritical density, a separation of starting materials, intermediates, or reaction products from the solvent CO_2 is easily achieved by simply changing temperature or pressure [5].

As early as 1978, the distinguished scientific journal *Angewandte Chemie* published a special issue covering the applications of $scCO_2$ in extractions and separations [6–10]. The work was pioneered by Kurt Zosel (Max-Planck-Institut für Kohlenforschung, Mülheim a. d. Ruhr, Germany), who first used $scCO_2$ to extract caffeine from roasted coffee beans, which is, to this day, the superior and standard industrial decaffeination procedure. Using Zosel's method, no residues of otherwise toxic extraction solvents are found (methylene chloride was used before this invention), leading to the high-quality and pure product that is consumed in large quantities today. Ever since the early 1970s, applications of $scCO_2$ as a preferred solvent have multiplied in many prominent areas, among which are food processing [11–13] and waste recycling [14–20]. Especially the latter area has experienced a substantial growth of applications in recent years.

In 1999, almost 20 years after the first special issue, the scientific journal *Chemical Reviews* dedicated yet another issue solely to supercritical media [21–24]. By comparing the two special issues, it becomes obvious that the focus of research had shifted from supercritical extractions and separations to chemical reactions and synthesis in supercritical fluids (SCFs). The shift reflects the increased significance of catalysis in modern chemistry and, in particular, shows an increased activity of catalysis research and applications in this field. Thereby, CO_2 is utilized in two different ways: foremost as a solvent in its liquid [25] or supercritical state, but also as a C_1 building block in chemical reactions. Frequently, even the combination of the two ways is found [26, 27]. As a C_1 chemical reagent, CO_2 is utilized to synthesize various indispensable base chemicals of the modern chemical industry such as carbonates [28, 29], urethanes [30, 31], lactones [32], methanol [33–36], as well as formic acid and its derivates [37–39]. Usually, these valuable products are formed industrially by catalytic hydrogenations (either homogeneous or heterogeneous) of carbon dioxide itself in the presence of suitable reaction partners such as alcohols or amines. The application of carbon dioxide as a liquid or a supercritical solvent is now widely established for extractions and separations (vide supra), hydrogenations and hydroformylations (vide infra), or for controlled polymerization reactions [40–42].

1.1 Supercritical Carbon Dioxide and Environmental Chemistry

To illustrate the relevance of SCFs (especially $scCO_2$) for the development of ecological and environmental processes, we examined the literature of academic and applied research regarding links between SCF research and environmental concerns, particularly within the last 12 years. As a result, the histogram in Fig. 1 shows the number of peer-reviewed articles and proceedings (data source SciFinder) published each year with a substantial emphasis on $scCO_2$ or environmental issues. Contributions to the literature were considered for the histogram in Fig. 1, if a connection to the terms "supercritical carbon dioxide" or "supercritical CO_2" or to "environmentally benign" or "environmentally friendly" was found in the title or abstract of the publication. In addition, a contribution was counted if it showed by conducting a keyword or content search for the terms listed before. In case of doubt of the substantial relevance for one or both of the subject fields, the article was examined more closely and decided upon separately.

Because of the close relationship between the pairs of terms analyzed (i.e., "supercritical carbon dioxide" with "supercritical CO_2" and "environmentally benign" with "environmentally friendly"), we further examined the

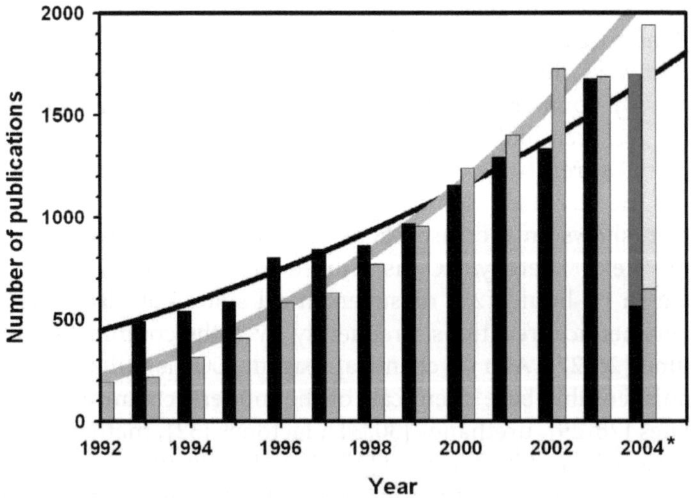

Fig. 1 Number of worldwide annual research and development publications (1992–2004) that contain the terms "supercritical carbon dioxide" or "supercritical CO_2" (*black bars*) or the terms "environmentally benign" or "environmentally friendly" (*gray bars*) in their title or abstract or that show up in a keyword or content search of those terms. Data for 2004 (*asterisk*) were derived from linearly extrapolating the number of entries from the first 4 months of 2004 (*darker part of the bars* for 2004). The *black and gray lines* depict the trend of the respective black and gray histogram data

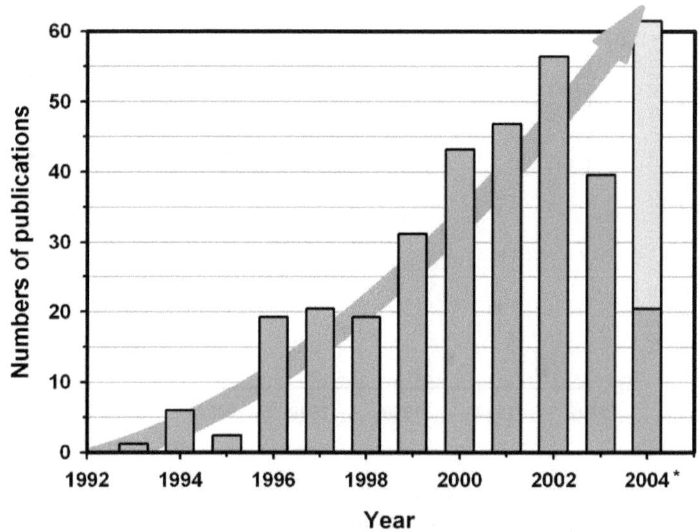

Fig. 2 Number of worldwide annual research and development publications (1992–2004) reporting environmentally motivated supercritical carbon dioxide ($scCO_2$) research. Data for 2004 (*asterisk*) were derived similar to those for Fig. 1. The *gray line* depicts the trend of the histogram data

entries to ensure that none of them were counted more than once in each of the categories. The histogram shows that the annual number of publications with emphasis on environmentally benign chemistry increases faster than the number of publications concerned with $scCO_2$ research. By combining the two groups of terms in our survey, we found that publications of environmentally motivated $scCO_2$ research are virtually nonexistent before 1992 but, ever since, their number has increased about proportional to the increase of environmentally motivated research and technology in general (Fig. 2).

As an additional finding, it is noteworthy that, within the $scCO_2$ research community, the number of environmentally related publications is continuously increasing by roughly 0.4% per year, reaching about 4.5% of all $scCO_2$ contributions in 2004 (Fig. 3). In view of that, a slower than originally expected but still steady increase of environmental focus in $scCO_2$ research is noted.

2
Physical Properties of Supercritical Fluids

Before we go into greater detail about potential applications of or in $scCO_2$, we will further discuss and illustrate the nature of an SCF. The supercriti-

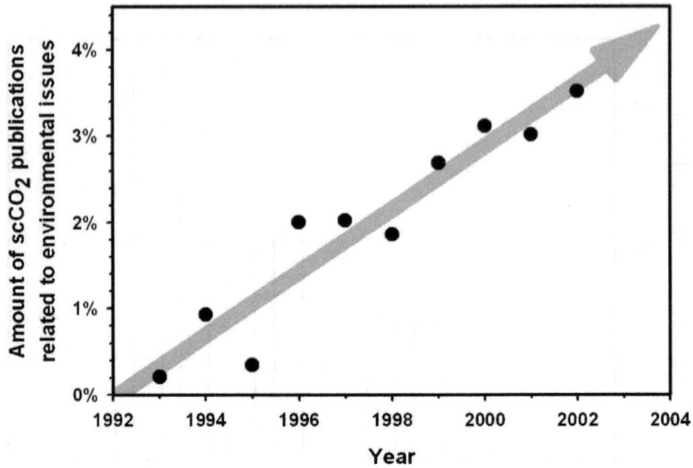

Fig. 3 Annual percentage of environmentally motivated publications with respect to all $scCO_2$ research and development publications. The *gray line* depicts the constantly increasing trend of the data

cal condition is often looked upon as a fourth state of matter in addition to the well-known gaseous, liquid, and solid states. In contrast to those "conventional" states, however, the borders to the supercritical range in a traditional phase diagram are not defined by phase transitions but only and precisely by the critical temperature, T_c, and the critical pressure, p_c. If both temperature and pressure of a pure substrate or a mixture exceed the critical values,T_c and p_c, the system becomes what is called supercritical. The special nature of an SCF becomes evident phenomenologically in that, above T_c, it will not liquefy no matter how far the pressure is increased. Likewise, above p_c, it will never transfer into a gas by increasing the temperature. Even though these effects rarely occur or pose a problem in daily life, they are quite well known. Butane, for example, with its critical temperature well above ambient temperatures (*n*-butane $T_c = 152\,°C, p_c = 37.96$ bar) is easily liquefied at room temperature by increasing the pressure and, thus, is commercially available as a liquid in pressurized tanks, whereas nitrogen does not liquefy this way. The critical temperature of nitrogen lies far below room temperature ($T_c =- 146.9\,°C, p_c = 33.94$ bar), leading to the inevitable conversion of nitrogen gas into an SCF upon reaching the critical pressure p_c but not to the condensation into a liquid. Consequently, gases with T_c below room temperature become liquefied under pressure only after lowering the temperature below T_c. In the commonly employed Linde process, the temperature reduction is achieved by exploiting the Joule–Thomson effect in gases.

For a compound or a mixture to become supercritical, no phase boundary is crossed regardless of starting from the liquid phase, the gas phase, or going along the liquid–vapor coexistence curve. The differences between liquid and gas disappear as soon as both T_c and p_c are exceeded. Using the SCF range (Fig. 4) as a detour, a substance may even be transformed from a liquid into a gas or vice versa without ever crossing a phase boundary. SCFs exhibit qualities of both liquids and gases, which are otherwise strictly separated states of matter. A transition between the solid state and an SCF also exists; however, it is usually not observed within the conventional ranges of pressure and temperature (Fig. 4).

Even though the range of an SCF is easily defined within a phase diagram, its very nature seems much more difficult to understand. It is hard to imagine that, above T_c, a system does not at all transfer into a condensed phase even at very high pressures, and the question remains what are actually the differences between a high-density SCF and a "true" liquid. Above p_c, the transition from a liquid to an SCF is conducted through increasing the temperature. If this transition is conducted with a system in a sealed container, it seems even harder to picture qualities changing for the overall density remains constant.

Accordingly, because SCFs can be similar to either gases or liquids or anything in between, their qualities are only suitably determined by bearing in mind their supercritical density. In Fig. 5, the density of CO_2 is shown as a function of pressure for various temperatures [43].

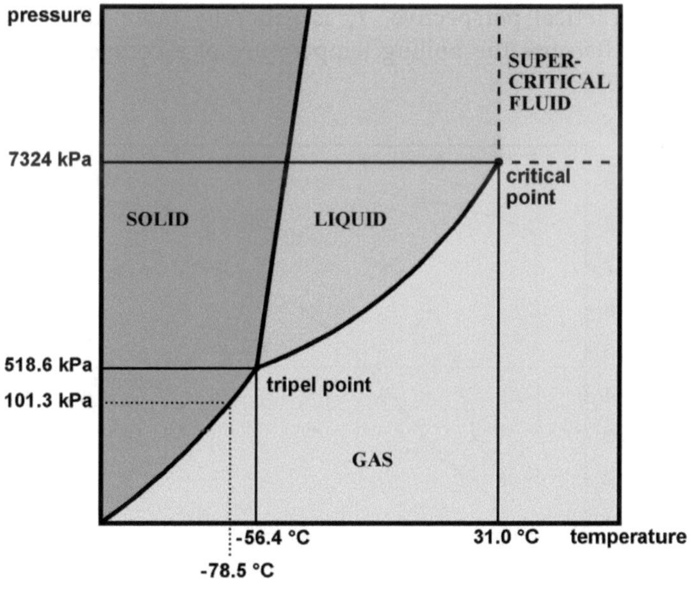

Fig. 4 Phase diagram of CO_2. Particularly highlighted are the critical point (T_c = 31.0 °C, p_c = 73.2 bar), the triple point (T_t =– 56.4 °C, p_t = 51.9 bar), and the sublimation boundary at ambient pressure (T =– 78.5 °C, p = 1.01 bar). Pressure values are denoted in kilopascals according to international standard units

By realizing that the density plays such an important role for the physical and thermodynamic properties of SCFs, the dissolution qualities of, and the transport phenomena in, $scCO_2$ may be deliberately tuned over wide ranges and adjusted for each individual application through temperature or pressure changes. The original and preferential use of $scCO_2$ in extractions and separations is largely based on the direct influence of the density on the dissolution power. For most applications in synthesis and catalysis, it is not merely sufficient to transfer CO_2 into an SCF but rather the density must be high enough to allow for the dissolution of one or more targeted compounds.

Besides CO_2, several other substrates are useful as supercritical solvents or reagents. Figure 6 shows various basic compounds arranged with respect to their critical data T_c and p_c. Because every pure compound and every mixture of compounds will eventually reach the supercritical range, only a selection of the most widely applied—and, hence, most relevant—substances is recorded. Moreover, gases with very low T_c are not listed, such as helium, neon, hydrogen, argon, nitrogen, or oxygen. One more preferential solvent for environmental and sustainable chemistry is water. However, water exhibits quite high values for T_c and p_c, is difficult to handle, and becomes very corrosive as an SCF. Still, several studies about reactions in supercritical water were published in recent years [45–48]. As apparent in Fig. 6, the critical values depend crucially on intermolecular forces such as dipole–dipole interactions, hydrogen bonding, and London dispersion forces (polarizability). From a practical perspective, T_c is generally more affected by these forces than p_c. Because the boiling temperature of a compound goes along

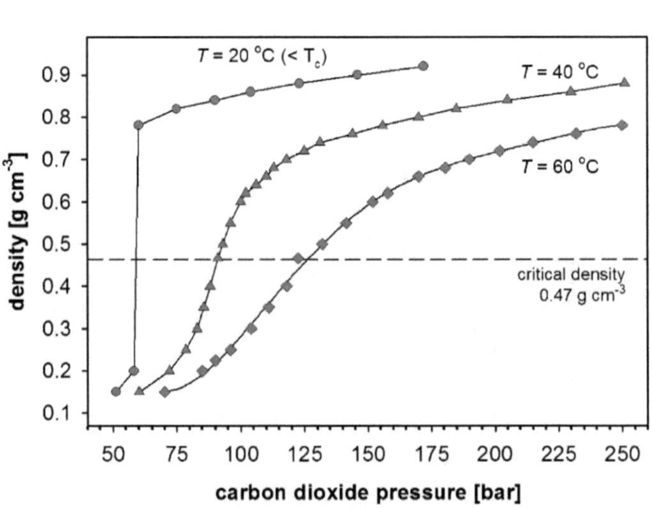

Fig. 5 Density as a function of CO_2 pressure at various temperatures [44]. Below the critical temperature T_c, a discontinuity occurs where CO_2 liquefies under the increase of pressure

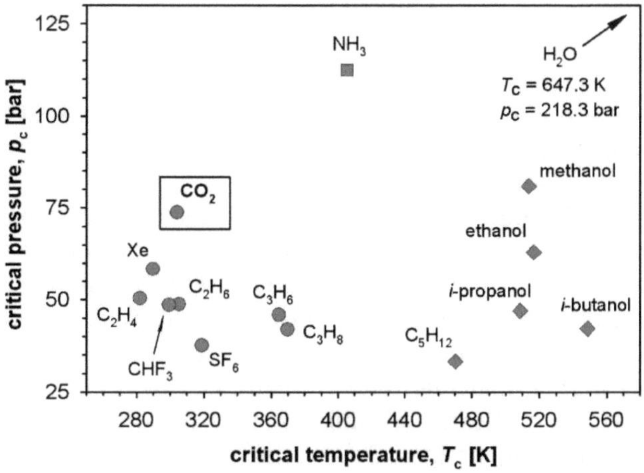

Fig. 6 Various base compounds arranged according to their critical values T_c and p_c. The values for CO_2 lie in a range of moderate pressure and comparably low temperature, while those for H_2O are beyond the area of display

with the quantity of intermolecular forces and the liquid–vapor coexistence curve shifts towards higher temperatures with increasing forces, T_c shifts accordingly. In water or alcohols, hydrogen bonding is the major source of intermolecular forces for which these compounds have critical temperatures that are often beyond practical and technical limitations.

2.1 Transport Phenomena in Supercritical Carbon Dioxide

2.1.1 Diffusion

In many, if not in the majority of, publications concerning research and applications of or in $scCO_2$, SCFs are referred to as media with dissolution capabilities of organic solvents while revealing the diffusivity of gases. Moreover, it is often stated that the combination of these particular qualities (i.e., dissolution capability and gaslike diffusion) engenders the high potential of SCFs for all kinds of applications. As pointed out before, however, density is the crucial parameter to reveal the properties of a SCF. Indeed, at densities about that of conventional organic solvents ($d > 0.5\,\mathrm{g\,cm^{-3}}$), $scCO_2$ exhibits the polarity and dissolution capability of typical nonpolar solvents such as hexane, benzene, or toluene. However, this density is approximately 4 orders of magnitude larger than that of a typical gas, and we entertained increasing doubts that diffusivities in dense $scCO_2$ reveal gaslike behavior.

Attending to our reservations, we measured diffusion coefficients using the newly developed NMR technique magnetization-grating rotating-frame imaging (MAGROFI) [49]. Diffusion coefficients determined by this technique are shown in Fig. 7 for various 1-alkenes dissolved in $scCO_2$ or toluene. The diffusion coefficients measured for toluene compared favorably to values obtained from the literature [50].

For toluene, all coefficients fall within one of the two hatched boxes of Fig. 7 depending on the two temperatures used for the experiments. It demonstrates that diffusivities in toluene increase by about a factor of 2 when the temperature is raised from 25 to 60 °C. In contrast, $scCO_2$ at 60 °C with a density similar to that of liquid toluene exhibits diffusivities that are about 5 times higher than the ones measured in toluene. Accordingly and even though $scCO_2$ and toluene exhibit similar dissolution capabilities, diffusivities are much higher in $scCO_2$ at the same temperature and density. Measurements conducted with other solvents (hexane or benzene) show similar behavior, namely that the diffusivities in $scCO_2$ are substantially higher. Consequently, chemical reactions conducted in $scCO_2$ are less likely to be limited by diffusion and, therefore, are usually much faster than in conventional organic solvents. However, the diffusivities of 1-alkenes in $scCO_2$ are still far from those observed in gases. In addition, it is noted that the diffusivities in $scCO_2$ are substantially influenced by the molecular mass of the alkenes [51], while this influence is less pronounced or almost negligible in conventional solvents.

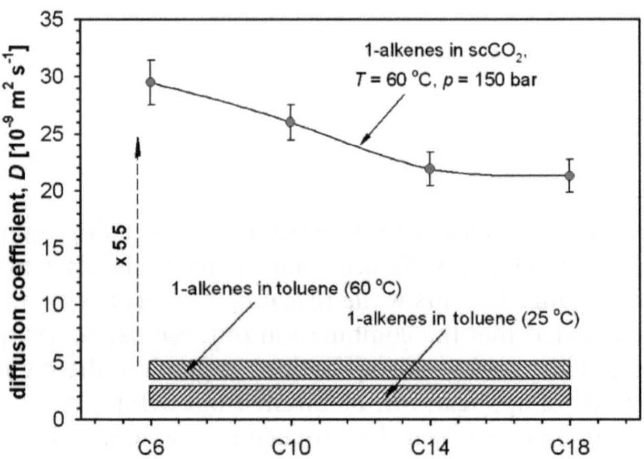

Fig. 7 Diffusion coefficients for 1-alkenes dissolved in $scCO_2$ or toluene as a function of the chain length of their carbon backbone. The labels *C6*, *C10*, *C14*, and *C18* on the abscissa denote the number of carbon atoms of the alkene, i.e., ranging from hexene (C6) to octadecene (C18) For the solvent toluene, all diffusion coefficients fall within one of the two *hatched areas* depending on the temperature. The diffusion in $scCO_2$, however, depends strongly on the length of the carbon backbone

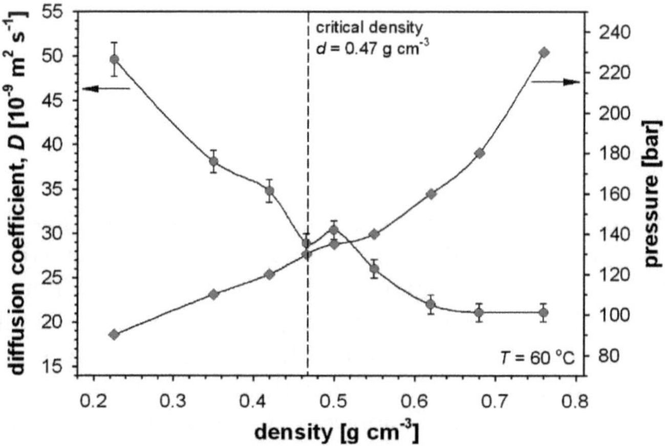

Fig. 8 Variation of the diffusion coefficient of 1-decene in $scCO_2$ as a function of density (*circles, left ordinate*). In addition, the pressure dependence of the density is shown (*diamonds, right ordinate*)

As mentioned before, the dissolution power of SCFs is tunable through variations in density. In addition, diffusivities of compounds dissolved in SCFs may be tuned as well provided that the density stays above the value that is needed for dissolving a compound. Figure 8 depicts how diffusion is tuned for 1-decene dissolved in $scCO_2$.

From the previous discussion, it is obvious that $scCO_2$ exhibits the dissolution capability of a conventional organic solvent if it is kept at a similar density. Because intermolecular interactions are much weaker than in standard solvents, compounds dissolved in $scCO_2$ show diffusivities enhanced by a factor of up to 1 order of magnitude. However, diffusion in gases is still up 4 more orders of magnitude, which is only reached in very low density $scCO_2$. Low-density $scCO_2$, however, has only marginal to no dissolution capability and, thus, loses all its potential as a solvent for chemical processes.

2.1.2 Viscosity

According to the findings just decribed, the enhanced reaction rates that have been reported for reactions in $scCO_2$ are not a special property of SCFs but simply occur because diffusion is enhanced in the low-viscosity medium CO_2. The low viscosity is the result of substantially reduced solvent–solvent molecular interactions in CO_2.

Diffusivity and viscosity in a liquid or a liquidlike SCF are physically interconnected as described by the equation of Stokes and Einstein:

$$\eta = \frac{kT}{6\pi r_A D}, \tag{1}$$

where η is the dynamic viscosity, k is Boltzmann's constant, T is the absolute temperature, r_A is the hydrodynamic radius of the diffusing particle, and D is the diffusion coefficient. Changing the diffusivity in SCFs will automatically change the viscosity, and both transport qualities are tuned simultaneously, yet not independently. From the finding that molecular diffusion in scCO_2 is typically enhanced up to 1 order of magnitude compared with the case for a conventional, nonpolar organic solvent (e.g., benzene), Eq. 1 calculates that, at the same temperature and density, the viscosity of scCO_2 is up to 1 order of magnitude lower [52]. While the reduced viscosity might not be as relevant for chemical processes in scCO_2, it can lead to important advantages in NMR spectroscopy in which signal linewidths are directly proportional to the sample's viscosity, and nuclear spin–spin relaxation is inversely proportional [53]. In certain investigations, especially, if quadrupolar nuclei are involved, the resolution of NMR spectra is substantially increased by the use of scCO_2 [54, 55]. In addition, the decreased nuclear spin–spin relaxation is utilized in highly resolved NMR imaging or diffusiometry experiments such as rotating-frame microscopy or MAGROFI, respectively [56, 57].

3 In Situ NMR Spectroscopy

Because parameters such as chemical shift, coupling constant, and relaxation time yield much information about chemical structure, reaction mechanism, and kinetics, NMR spectroscopy is often a desirable analytical tool for in situ investigations of chemical reactions. For example, homogeneously catalyzed hydrogenations, hydroformylations, or polymerizations are frequently examined with in situ NMR [58]. As discussed in the previous chapter, the usefulness of in situ NMR for investigations in SCFs, however, depends crucially on effective density control capabilities in the available NMR probes. Put another way, in situ NMR in SCFs relies on the capability to monitor and most accurately adjust temperature and pressure. While temperatures ranging from -70 to 180 °C are conveniently and safely achieved with standard high-resolution NMR probes by air-stream heating or cooling, enhanced pressure capabilities are more problematic. Large amounts of energy are stored in NMR sample containers if in situ investigations are conducted with SCFs or with reactions that proceed under high gas pressure. Yet ample gas pressure ranges are essential for kinetics investigations, e.g., for studies of rate constants in gas–liquid reactions in which the amount of dissolved gas depends on the partial pressure above the liquid [59, 60].

In the literature, several pressure-vessel NMR probes are described, which are especially designed for high-resolution in situ investigations of chemical reactions under high gas pressure [61–63]. Alternatively, single-crystal sapphire cells are used as high-pressure sample tubes [64, 65]. The major advantage of the latter is that no modification of standard NMR probes is needed. For safety reasons, however, we prefer high-tensile-strength metal over single-crystal material in high-gas-pressure applications. Especially, small cracks or hairline fractures that quite frequently happen to the commercially available sapphire tubes bear the danger of explosive destruction. In metal vessels, however, these fractures are less likely to occur and, in addition, will release the pressure in a rather controlled manner.

3.1 Toroid Cavity Autoclave Probes

Because of our safety concerns with high-pressure NMR probes but also because of sensitivity and resolution considerations, we will review the recent toroid cavity autoclave (TCA) concept that has been applied very successfully to high-resolution NMR under high gas pressures or in SCFs. The concept utilizes toroid cavity detectors (TCD), i.e., coaxial NMR resonators that have previously been mounted only inside cylindrical metal autoclaves. For example, micelle formation in $scCO_2$ was investigated this way [66] or pressure-dependent spin-rotation relaxation of gaseous methane and hydrogen [67, 68], thereby exceeding the pressure range of previous studies by an order of magnitude. In addition to the high-pressure applications, TCDs are known for their capabilities of rotating-frame microscopy [69–71], diffusiometry [72, 73], and relaxometry [74, 75] under various conditions.

As a coaxial cavity resonator, TCDs consist effectively of a central conductor and an outer cylindrical shielding, which both exhibit the axial symmetry of a standard cylindrical autoclave. Because of this common symmetry, the combination of a TCD and an autoclave in one structural part of an NMR probe leads to the TCA [76]. The fundamental idea behind this concept is to seal a TCD to the outside and increase its outer-wall thickness, so that it withstands high gas pressures. Figure 9 shows two different types of TCAs, one built preferably for SCFs (Fig. 9a) and the other one for gas–liquid reactions (Fig. 9b). TCAs can be machined to fit the limited space of narrow-bore superconducting NMR magnets in which the free space for probes is only 40 mm in diameter.

Four important issues must be addressed when an appropriate material is chosen for constructing TCAs: (1) low absolute magnetic susceptibility, (2) high tensile strength, (3) high electrical (and thermal) conductivity, and (4) chemical inertness. While issues 1, 2, and 3 practically limit the num-

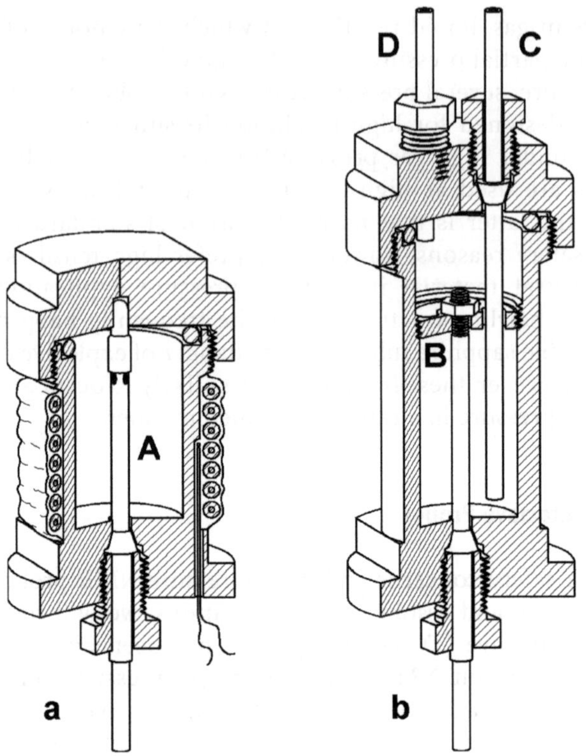

Fig. 9 First-generation of homebuilt toroid cavity autoclaves (*TCA*) for in situ NMR investigations of **a** supercritical fluids *SCFs* or *b* gas–liquid reactions under high-pressure and, simultaneously, high-temperature conditions [76]. The pressure capillary of the TCA for SCFs (*A*) simultaneously functions as the central conductor for the toroid NMR resonator. While the entire sample volume in the TCA for SCFs (**a**) is sensitive for NMR experiments, only the lower part in the gas–liquid TCA (**b**) is used to receive NMR signals. The gas–liquid TCA is charged with the reactive solution to above the intermediate contact disk (*B*), so that no susceptibility mismatch from the gas–liquid interface disturbs the magnetic field lines in the NMR-sensitive area. Two pressure capillaries are used to vigorously inject reactive gases (capillary *C*) and to recycle, or dispose of, excess gas that leaves the reaction solution (capillary *D*)

ber of materials available and useful for TCAs to copper-based alloys such as diamagnetic phosphorous bronze (P-bronze) or beryllco alloy (beryllium–copper alloy), issue 4 necessitates a special treatment of the interface between the sample and the TCA body. The absolute susceptibilities of other well-known high-tensile-strength materials for autoclave construction (e.g., brass, stainless steel, or titanium–aluminum alloy) are too high to qualify for TCAs. In addition, for highly sensitive NMR experiments in TCAs, it is important that the autoclave material, which also serves as the NMR resonator, exhibits maximum electrical conductivity. This inevitably leads to the same copper-based alloys mentioned before (P-bronze or beryllco alloy). Advantageously, these particular alloys also offer a high thermal conductivity for efficient temperature control and adjustment. Most of our TCAs were, therefore, machined from commercially available P-bronze that has a diamagnetic susceptibility close to that of organic solvents or water [77]. It has a tensile strength of 690 MPa at room temperature, which drops off at higher temperatures but is still satisfactory for most SCF applications. The inner walls of the TCA and the central conductor are usually gold-plated electrolytically, so that no undesirable chemical reaction occurs at the interface between the TCA material and the sample under investigation. A typical undesirable reaction in rhodium–organic or platinum–organic catalysis, for example, is the formation of rhodium black or platinum metal [i.e., Rh(0) or Pt(0), respectively], which occurs if the organometallic complexes interact with Cu(0) from the P-bronze. A gas, liquid, or SCF distribution system is connected to the TCA by (polyether)ether ketone (PEEK) capillaries or, if higher pressure limits are demanded, by stainless steel or other high-tensile-strength metal capillaries. PEEK capillaries are usually preferred over electrically conducting metal capillaries because they do not introduce external noise into the radiofrequency resonance circuit of the probe. For further technical details about the design of TCAs, we refer the reader to the literature [76].

A prominent feature of TCAs is that the sample volume is completely sensitive for NMR measurements, so the highest possible filling factor is achieved and no volume element exists which is not penetrated by the radiofrequency field of NMR pulses (B_1). Thus, a high sensitivity is achieved with a minimum amount of sample. Because of the minimized sample volume, the energy that is stored in the TCA for high-pressure experiments is also kept to a minimum. Recently, an advanced TCA design was developed, which is optimized for studies in which reactions in SCFs are directly compared with the same reactions in liquid solutions and under the same gas pressure [78]. This sensitive high-resolution probe (Fig. 10) is equipped with a nonreturn valve through which gaseous or liquid components can be injected even if high pressure has already been applied to the reactor. In addition, a coaxial resistive heating arrangement was added, which, although electrically operated, does not disturb the main magnetic field (B_0) of the NMR magnet. With these additional features, the new TCA is particularly useful for kinetics investigations in which a defined starting point of the reaction is desirable and a high degree of temperature stability is indispensable.

The new probe combines key features of both previously described TCAs in one probe and, at the same time, improves the pressure, temperature, and imaging capabilities. Its design is modular, so that the probe and the autoclave are effortlessly dismantled, and all parts are easily accessed. Hence, the autoclave and its periphery can readily be cleaned thoroughly and single parts may be replaced without rebuilding the entire TCA. The modular approach is particularly beneficial for future research projects in which ad-

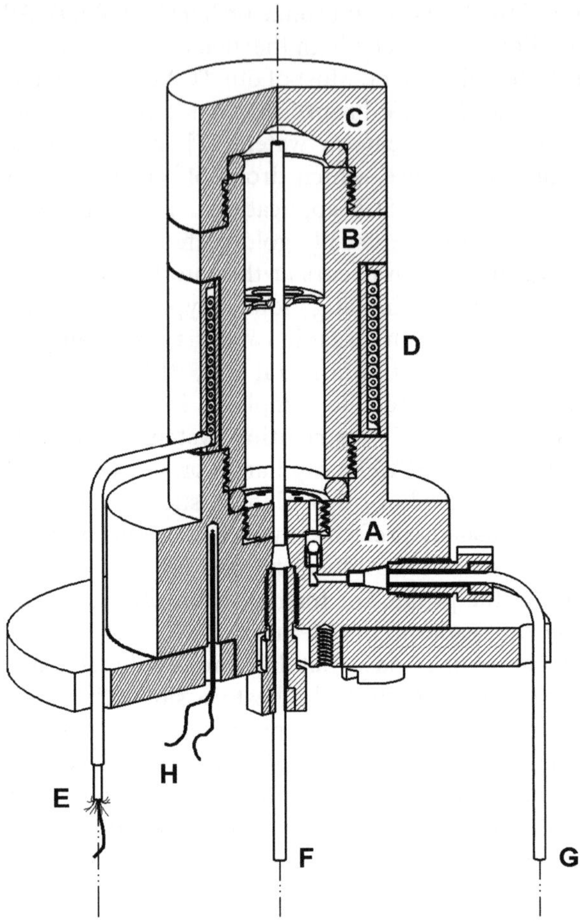

Fig. 10 Advanced TCA design that combines the different types of Fig. 9 into one NMR probe [78]. It is built particularly for studies in which catalytic reactions in SCFs are directly compared with the same reactions in liquid solutions under the same gas pressure. The metal pressure vessel comprises a base (*A*), a body (*B*), and a cap (*C*). It is surrounded by a cylindrical housing (*D*) that tightly encloses a coaxial, resistive heater (*E*) which does not induce spurious magnetic fields. Two pressure capillaries lead into the pressure vessel, where one (*F*) simultaneously functions as the central conductor of the toroid NMR resonator; the other one (*G*) is used to inject key reagents that initiate the catalytic reaction under investigation. The temperature is monitored by a copper-constantan thermocouple wire (*H*)

ditional capabilities are desired or in which additional peripheral equipment will be required.

TCA resonators are usually double-tuned to a high frequency (^{1}H or ^{19}F) between 200 and 400 MHz depending on the magnetic field strength B_0, and a lower frequency (e.g., ^{2}H) to provide for a field-frequency lock. Alterna-

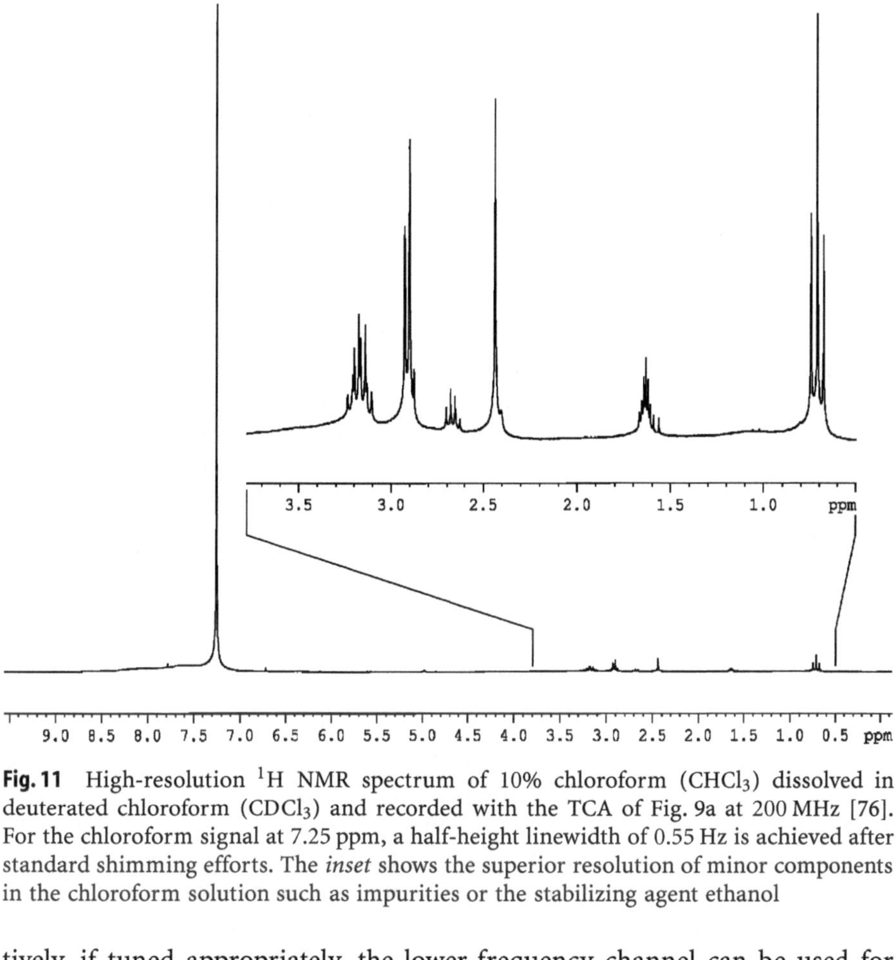

Fig. 11 High-resolution ^1H NMR spectrum of 10% chloroform ($CHCl_3$) dissolved in deuterated chloroform ($CDCl_3$) and recorded with the TCA of Fig. 9a at 200 MHz [76]. For the chloroform signal at 7.25 ppm, a half-height linewidth of 0.55 Hz is achieved after standard shimming efforts. The *inset* shows the superior resolution of minor components in the chloroform solution such as impurities or the stabilizing agent ethanol

tively, if tuned appropriately, the lower-frequency channel can be used for investigations with NMR-active nuclei such as ^{13}C, ^{31}P, ^{27}Al, ^{29}Si, or ^{195}Pt. Signal-to-noise ratios (SNR) measured with TCA probes from standard solutions such as 0.1% ethyl benzene and 1% tetramethysilane in deuterated chloroform (e.g., SNR = 95 at 200 MHz) compare favorably to those of standard 5-mm high-resolution NMR probes (e.g., SNR = 85 at 200 MHz). Even the spectral resolution is comparable to that of standard NMR probes, although the TCA is fabricated from a large piece of metal that is in close contact with the sample [76]. Figure 11 shows the ^1H NMR spectrum of a solution of 10% chloroform ($CHCl_3$) in deuterated chloroform ($CDCl_3$) recorded with a TCA at 200 MHz. A half-height linewidth of 0.55 Hz is achieved for the chloroform signal at 7.25 ppm after regular shim efforts. It is noted, however, that the signal bases are often considerably distorted, i.e., they are broadened to higher frequencies by about 200 Hz. The amount of distorted signal is usually reasonably small and does not normally hinder in situ NMR investigations.

3.2 Parahydrogen-Induced Polarization

Although NMR is a superior tool for in situ investigations of chemical reactions, it is inherently insensitive when compared with spectroscopic methods such as IR or UV/vis spectroscopy. The magnetization and, thus, the NMR signal intensity that is yielded from nuclear spins at thermal equilibrium in the magnetic field of a typical NMR spectrometer is only 10^{-5} to 10^{-4} of what it could be if all spins were aligned in parallel. To enhance the intrinsically small NMR signal amplitudes, several hyperpolarization strategies have been invented creating large non-Boltzmann (i.e., nonequilibrium) magnetizations. Among those is the parahydrogen-induced polarization (PHIP) technique that utilizes the spin alignment of parahydrogen (p-H_2), i.e., the nuclear singlet state of dihydrogen, to hyperpolarize intermediates and products of hydrogenations. We will shortly review this technique; yet, for further details, we recommend the reader consult the many review articles that are available emphasizing the different methodical, practical, or chemical aspects of PHIP [79–86].

Molecular hydrogen is enriched in its nuclear singlet state (i.e., enriched in p-H_2) by passing gaseous H_2 at ambient or slightly elevated pressure (20 bar or less) over gas-adsorbing, paramagnetic materials that are kept at liquid nitrogen temperature or below. To conduct this process, homebuilt [82] or commercial [87] apparatuses are available, which provide for continuous streams of p-H_2-enriched gas by utilizing coarse-grained activated charcoal as the paramagnetic material. Hydrogenations conducted with the p-H_2-enriched gas stream can induce large nuclear spin hyperpolarization in the intermediates or products of the reaction, which, in turn, can lead to greatly enhanced NMR signals. The prerequisites for observing spin hyperpolarization in the NMR spectra acquired during or after hydrogenation are:

(a) The two hydrogen atoms must be transferred in pairs.
(b) The nuclear spin relaxation of the transferred hydrogen atoms must be slower than the overall kinetics of the chemical reaction.
(c) The symmetry of the transferred hydrogen atoms must be broken during the reaction.

Conditions a and b allow for the nuclear singlet information of p-H_2 to remain—at least partially—intact during the reaction and, together with the hydrogen atoms, become transferred to the reaction intermediates and products. Consequently, the product's energy levels with a nuclear singlet character for the transferred hydrogen atoms ($\alpha\beta$ and $\beta\alpha$ in the example of Fig. 12) are exclusively—or, at least, more heavily—populated than others. Condition c is necessary to obtain NMR-observable magnetization from the hyperpolarization. Spectra recorded in situ during, or shortly after, the reaction with p-H_2 exhibit greatly enhanced NMR signals in unique absorption and emission patterns (Fig. 12). These patterns reflect the fate of the hydro-

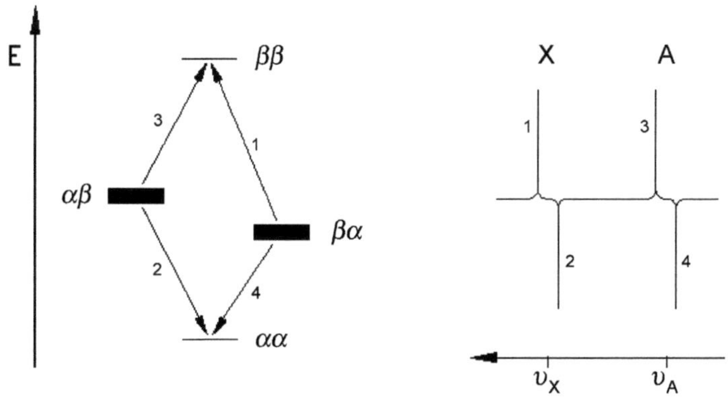

Fig. 12 Population of the spin energy levels (*left*) of a hydrogenation product with AX spin system for the transferred hydrogen atoms after an instantaneous reaction with pure parahydrogen p-H_2 [82]. The energy axis is labeled with the letter *E*. Only energy levels with nuclear singlet character ($\alpha\beta$ and $\beta\alpha$) are populated. The schematic spectrum (*right*) depicts the greatly enhanced NMR signals in absorption and emission as expected if the spectrum is recorded in situ during the reaction with p-H_2. The frequency axis is labeled with the resonance frequencies ν_A and ν_X of the A and X nuclei, respectively, where ν increases from right to left according to NMR convention

gen atoms during a chemical reaction or catalytic cycle. With the help of pulse sequences and experimental procedures, they are utilized to elucidate mechanisms and kinetics of hydrogenations and to identify reactive intermediates that are not seen with other methods. Similar but inverted NMR signal patterns are achieved with H_2 enriched in its nuclear triplet state (i.e., enriched in orthohydrogen) [88]. Sometimes, polarization signals are even observed if regular H_2 is used, i.e., H_2 that is neither enriched in parahydrogen nor in orthohydrogen but is left at its thermal equilibrium [89]. Thermal H_2 hyperpolarization, however, reveals much smaller signal enhancements.

Because of the lower viscosity, NMR relaxation in scCO_2 is slower compared with the relaxation in conventional solvents. As a consequence, PHIP signals decay much slower in experiments with scCO_2 than in experiments with standard solvents. This advantage, however, is sometimes counteracted by very high reaction rates in scCO_2, which render it much more difficult to observe the kinetics of a reaction quantitatively, especially, if evaluating the starting kinetics. While this difficulty has been addressed by incorporating a second pressure line for injecting key ingredients through a nonreturn valve into the TCA (Fig. 12), high reaction rates also lead to large quantities of product, whose standard non-PHIP NMR signals may interfere with the quantitative evaluation of the PHIP signals. In Fig. 13, a typical PHIP spectrum is shown from the—moderately fast—homogeneous hydrogenation of a fluorinated acrylic acid ester in scCO_2 (cf. Sect. 5.2).

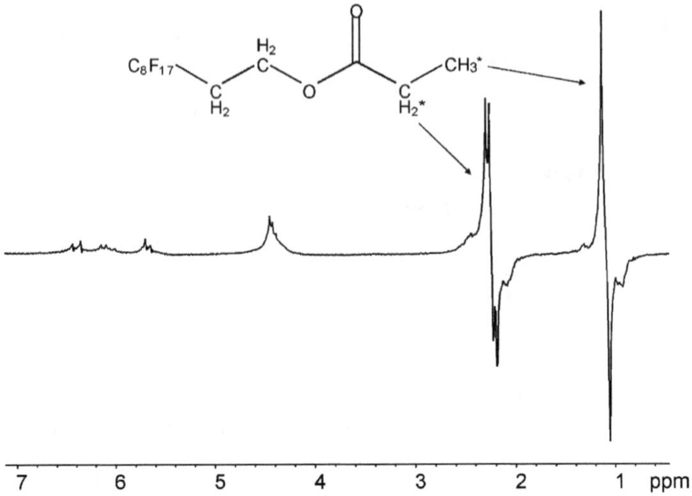

Fig. 13 p-H_2-induced polarization (*PHIP*) NMR spectrum acquired during the homogeneous hydrogenation of a fluorinated acrylic acid ester with p-H_2 in $scCO_2$ [$p(CO_2) = 90$ bar, $p(H_2) = 10$ bar, $T = 50\,°C$]. $H_2Ru(PPh_3)_4$ was used as the catalyst [76]. The *asterisks* in the chemical structure of the fluorinated ester mark the position in which the hydrogen atoms were transferred

4 Homogeneous Catalysis

Very diverse concepts for catalysis in SCFs were developed in recent years, ranging from purely heterogeneous systems [90–92] to various categories of homogeneous catalysis including biochemical and enzymatic reactions [93–95]. While a full review of all concepts would be beyond the scope of this article and, at best, would limit the description to a rather sketchy and superficial treatment, we will exemplify in detail homogeneous and colloidal catalysis in $scCO_2$. Among those, we will focus on transition-metal complexes of rhodium, ruthenium, and palladium, which exhibit the highest potential for specialized applications in SCFs. Other important transition metals for SCF catalysis are cobalt [96–98] and manganese [99]; however, they are less frequently used nowadays.

4.1 Hydroformylation

Among the first catalytic reactions conducted and investigated in SCFs were hydroformylations through which aldehydes are formed by the simultaneous transfer of H_2 and carbon monoxide to alkenes. Hydroformylations are, together with hydrogenations, the most important reactions that involve the transfer of hydrogen. They are particularly important in industrial synthesis, because a C_1 unit is added, lengthening the carbon backbone of the alkene while revealing a reactive oxidation state intermediate between alcohol and carbonic acid [100–103]. Aldehydes of almost any kind are obtained from hydroformylations and are readily converted into a vast variety of fine chemicals, such as carbonic acids (via oxidation) or long-chain alcohols (via aldol coupling). In the chemical industry, however, the majority of hydroformylations involve the starting materials propene or butene.

It was in the late 1980s when Rathke et al. [96, 104, 105] started their pioneering work with hydroformylations in $scCO_2$. Their work appears to be the first systematic studies of catalysis in SCFs, thereby utilizing the standard industrial catalyst dicobalt octacarbonyl for converting propene to butane aldehyde. Although $scCO_2$ is still the most frequently used SCF for catalysis, more recently, several other SCFs were successfully employed as media for hydroformylation reactions [106–110]. Nowadays, SCF hydroformylations are also conducted as part of biphasic reactions and, particularly, in combination with ionic liquids [111, 112]. Figure 14 shows the structures of the two building blocks (A and B) for a modern rhodium-organic catalyst that has been optimized for the homogeneous catalytic hydroformylation of norbornadiene or 1-decene in $scCO_2$. The catalytically active complex is formed in situ during the reaction.

Most often, catalyst precursors for hydroformylation reactions are carbonyl complexes of the transition metals cobalt [96–98, 104, 105], ruthenium [114], rhodium, or manganese [99, 115]. In very many cases, efficient hydroformylation catalysts are less effective or even ineffective for hydrogenations. This is commonly explained with the difference in hydrogen transfer mechanisms between hydrogenations and hydroformylations. While homogeneous hydrogenations proceed mainly by the pairwise transfer of hydrogen atoms, hydroformylations take place by transferring a hydride ion, i.e., proceed by an ionic mechanism that may be assisted by a base. Because of this ionic mechanism in which a single hydrogen atom is transferred, studying hydroformylations with PHIP NMR spectroscopy (vide supra) remains a difficult task. PHIP relies on the preservation of the H_2 nuclear singlet information, which is easiest achieved by a pairwise transfer of the hydrogen atoms. Still, it has been proposed [116] and recently verified experimentally [117] that single hydrogen atom transfer may lead to similarly enhanced PHIP NMR signals.

4.2 Hydrogenation

Homogeneously catalyzed hydrogenations in $scCO_2$ were first conducted during the early 1990s by Jessop and coworkers [21, 118, 119], who hydrogenated

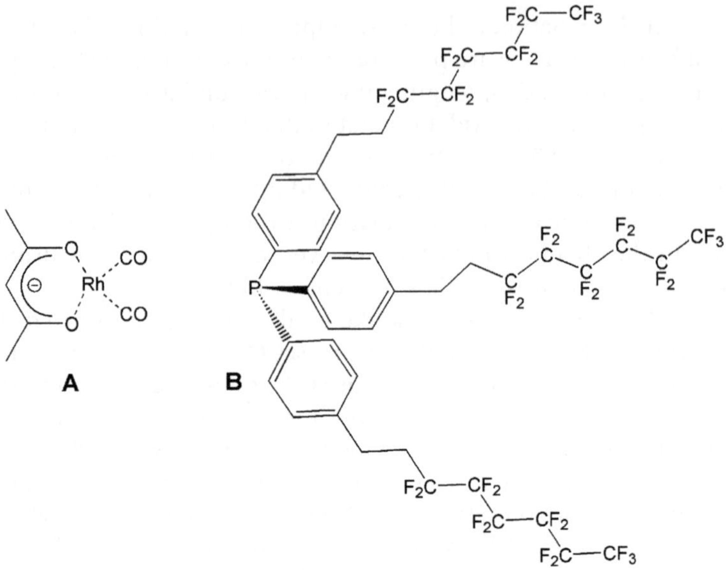

Fig. 14 Schematic structures of the building blocks for a rhodium–organic catalyst (**a** catalyst precursor, **b** ligand) optimized for the homogeneous hydroformylation of norbornadiene or 1-decene in $scCO_2$ [113]. The catalytically active complex forms in situ during the reaction

CO_2 itself to form formic acid and its esters. Like other research groups in the following years [120, 121], they exploited the knowledge about catalytic hydrogenations in conventional solvents and decided to use neutral ruthenium complexes as the catalyst. This type of catalyst dissolves easily in nonpolar organic solvents and, thus, was considered a good candidate for homogeneous catalysis in SCF. More recently, similarly structured neutral ruthenium complexes were also used for hydroformylation reactions in $scCO_2$; however, they exhibited a substantially reduced activity compared with that of rhodium complexes [122].

In addition to neutral transition-metal complexes, positively charged, cationic complexes are widely used as catalysts for homogeneous hydrogenations and hydroformylations. Especially in rhodium–organic catalysis, the cationic complexes are more selective (i.e., regio-, stereo-, and enantioselective) than their neutral counterparts. However, cationic complexes dissolve only poorly in nonpolar solvents like $scCO_2$. To overcome this limitation and to utilize the extra selectivity, $scCO_2$-philic perfluorinated side chains such as C_6F_{13} moieties (so-called pony tails) were introduced into the ligands of transition-metal complexes to substantially improve the $scCO_2$ solubility of cationic catalysts [123–129]. Alteration of the structure of a highly refined ligand, however, may alter the catalyst's electronic properties and, consequently, may impact the all too often most particularly engineered catalytic activity and selectivity. Therefore and as a more general concept, the pony tails were introduced spaced away by additional short alkyl chains (i.e., C_2H_4), so that the dissolution properties of the fluorinated side chains are isolated from the ligand framework. The utilization of alkyl "spacers", however, imposes an even larger synthetic challenge than the introduction of perfluorinated pony tails alone, making the ligand once more the most expensive part of homogeneously catalyzed hydrogenations. A typical rhodium–organic catalyst with pony tails and spacers is shown in Fig. 15 as designed and synthesized by Leitner et al. [125] particularly for the stereoselective hydrogenation of itaconic acid esters in $scCO_2$. This most active and $scCO_2$-philic catalyst is formed by a ligand-exchange reaction of a suitable rhodium complex with the specialized pony-tail ligand. Additionally, the fluorinated anion tetrakis(3,5-bis(trifluoromethyl)phenyl) borate (BARF) was chosen to compensate the positive charge and to increase the solubility in $scCO_2$ even further [130, 131].

With the catalyst of Fig. 15, enantiomeric excesses of about 75% were yielded for the *R* product. In situ PHIP experiments conducted with p-H_2 revealed NMR spectra (Fig. 16) that were virtually identical to those obtained from the same reaction in conventional solvents such as toluene or THF. Consequently, the catalytic cycle in $scCO_2$ proceeds almost identically to the reaction mechanism in standard solvents, and a pairwise vicinal addition of H_2 prevails [125]. The reactions in conventional solvents were conducted with

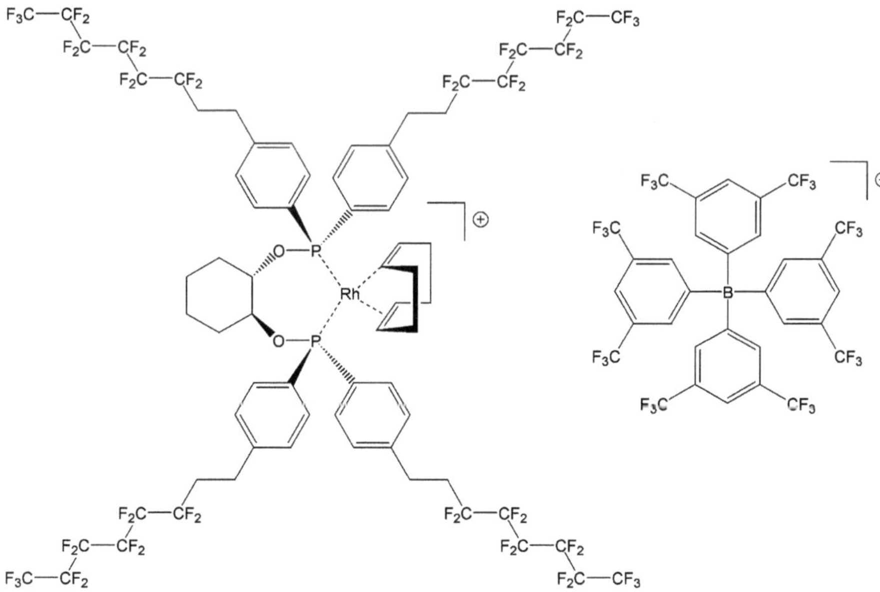

Fig. 15 Schematic structure of a rhodium–organic catalyst with "pony tails" (C_6F_{13}) and "spacers" (C_2H_4) as synthesized by Leitner et al. [125] for the stereoselective hydrogenation of itaconic acid esters in $scCO_2$

an unsubstituted rhodium–organic catalyst, i.e., with a catalyst equivalent to the one used in $scCO_2$ but without the spacers and pony tails. While a similar and sometimes even enhanced selectivity is observed for catalytic hydrogenations in $scCO_2$, substantially increased reactivity is often detected [132–134].

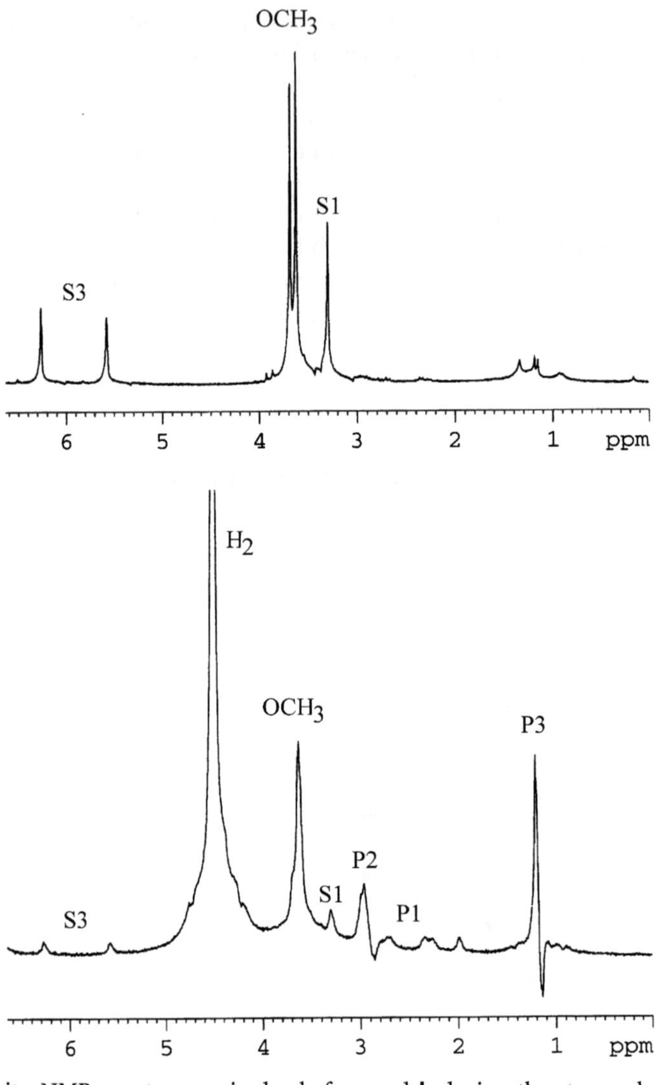

Fig. 16 In situ NMR spectra acquired **a** before and **b** during the stereoselective hydrogenation of itaconic acid ester using p-H_2 and the rhodium–organic catalyst shown in Fig. 15 [125]. The labels *S* and *P* mark signals originating from the reagent and the reaction product, respectively. In **b**, PHIP signals from the reaction product are noticed, which are identical to those detected from the same reaction in conventional solvents. Because of the fast reaction in $scCO_2$, almost all of the reagent is already consumed

From the mechanistic investigations with PHIP, which revealed identical mechanisms, it is evident that these differences are due to differences in the physicochemical properties of $scCO_2$ (e.g., in the high gas miscibility or the enhanced diffusivity) or to subtle influences that CO_2 might have on the relative rates or equilibria within the catalytic cycle.

Besides improving a catalyst's solubility through modifying its ligand with pony tails or utilizing the particularly soluble anion BARF, the reagent that is to be hydrogenated may also enhance the solubility and reactivity of a catalyst. Because an intermediate catalyst–reagent complex is necessarily formed during the catalytic cycle, the catalyst's solubility is automatically increased if the reagent is highly soluble. The reagent is usually available in large quantities and, therefore, acts as a cosolvent for the catalyst. The increased solubility of the catalyst because of the reagent's cosolvent effect can lead to substantially increased reaction rates, as studied in detail by quantitative in situ NMR [57]. In contrast, however, the reactivity of a catalyst might substantially decrease as a fluorinated reagent is consumed, and the cosolvent effect disappears. Figure 17 shows NMR spectra obtained from the hydrogenation of either styrene or 3-trifluoromethyl styrene with the nonfluorinated ruthenium complex $H_2Ru(PPh_3)_4$ in $scCO_2$.

Both reagents were hydrogenated in a TCA for 20 min under the same supercritical conditions for the solvent CO_2 [$p(CO_2)$ = 85 bar, $T = 50\,°C$]. In addition, the same concentrations of catalyst and reagent were applied for both reactions, resulting in an identical substrate-to-catalyst ratio (SCR). Merely the hydrogen pressure, $p(H_2)$, was different in the two experiments that led to the spectra in Fig. 17. However, even though $p(H_2)$ was twice as high for the hydrogenation of the unsubstituted styrene, the conversion of the fluorinated substrate 3-trifluoromethyl styrene proceeded much faster. Through control experiments with lower hydrogen pressures, it was verified that the higher $p(H_2)$ did not impose an inhibiting effect on the styrene catalysis. Accordingly, the introduction of a simple CF_3 moiety into the reagent molecule (the *meta* position was chosen for the least electronic influence on the vinyl group) enhances the cosolvent effect for the catalyst, so the turnover frequency (TOF) increases substantially. In addition, an undesired side reaction that is observed for the reagent styrene, i.e., the formation of polymerized product, is not detected with the reagent 3-trifluoromethyl styrene.

4.2.1 Colloid Catalysis

Besides complexes with a single transition-metal atom at the center, transition-metal clusters and colloids are used for homogeneous catalysis. The colloid catalysts are typically transition-metal particles that range from less than 100 atoms up to a few 1000 atoms. Their qualities are intermediate be-

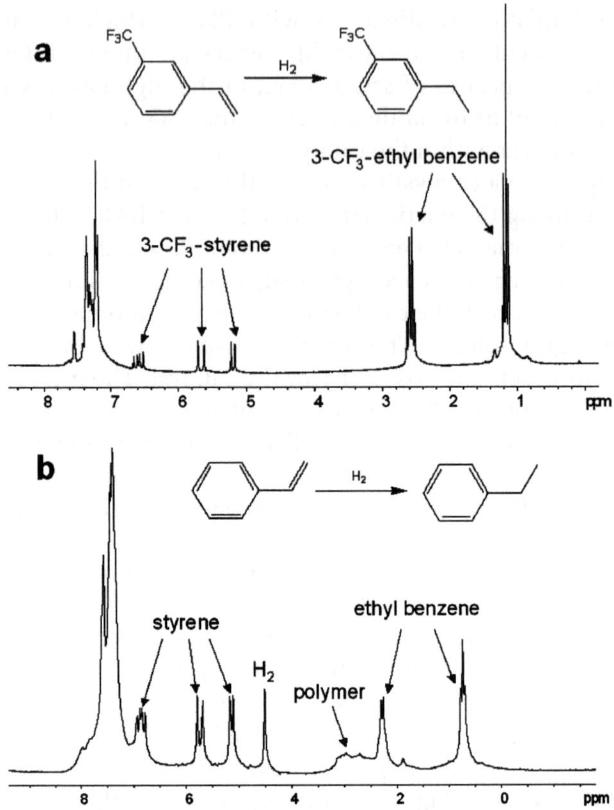

Fig. 17 NMR spectra recorded after the hydrogenation of **a** styrene [$p(H_2) = 40$ bar] and **b** 3-trifluoromethyl styrene [$p(H_2) = 20$ bar, $T = 50$ °C, reaction time 20 min] using the nonfluorinated ruthenium complex $H_2Ru(PPh_3)_4$ as the catalyst in $scCO_2$ [$p(CO_2) =$ 85 bar, $T = 50$ °C, reaction time 20 min]. Even with only half the pressure of H_2, the hydrogenation of 3-trifluoromethyl styrene (**a**) proceeds much faster. In addition, no polymeric by-product is detected as in **b**

tween those of metal clusters that typically contain only between two and ten transition-metal atoms and metallic surfaces. Hence, colloid catalysts are often envisioned as transitional between homogeneous and heterogeneous catalysis. In conventional solvents, metal colloids have been applied for various types of catalytic reactions [135–138], and their tunable properties have been studied quite intensely [139–142]. Only little, however, is known about applications of these colloids in SCF catalysis, even though the particle size and the tunable qualities of colloids particularly augment the tunable properties of SCFs. Subsequent to a catalytic reaction, for example, the solvent CO_2 is easily removed quantitatively by venting the system (i.e., by reducing the pressure below p_c), and the only task left is separating the catalyst from the reaction products. For colloid catalysts, this is easily achieved quantitatively by ultrafiltration [143–145] or vacuum distillation [146, 147]. Accordingly, reaction products such as food ingredients or pharmaceuticals, which require a high purity and, particularly, should be free of even traces of toxic organic solvents, would largely benefit from the combination of colloid catalysts and SCF.

The applicability of colloid catalysts in SCFs was first tested by Zemaian et al. [148] and yielded only very low conversion rates (less than 1%). These early investigations suggested that the potential for colloid catalysis in SCFs is quite limited, mainly because of the poor solubility of the colloids. Recently, however, the limited solubility has been addressed by using inverted micelles [149, 150], emulsions [151, 152], or biphasic systems [153, 154]. In the biphasic systems, colloid catalysts are dissolved in a polar solvent such as water [153], different alcohols, or ionic liquids [155–159], while the reagents dissolve in nonpolar media such as $scCO_2$, liquid CO_2, or ethane [160, 161]. In all of these cases, reaction products are obtained with high purity, and the catalyst is most readily recycled. However, the catalytic activity in biphasic systems is still somewhat limited because of the slow but necessary phase transfer of the reagent from the nonpolar to the polar phase and the equiva-

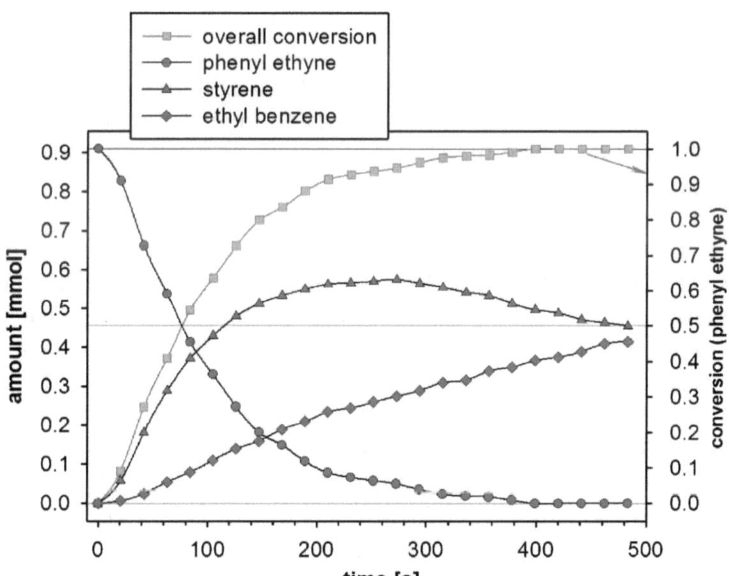

Fig. 18 Amount (*left ordinate*) of reagent (*circles*), intermediate product (*triangles*), and final reaction products (*diamonds*) yielded from the colloid-catalyzed hydrogenation of phenyl ethyne to styrene and, subsequently, from styrene to ethyl benzene as a function of reaction time [44]. In addition, the conversion (*light gray squares*, *right ordinate*) of the starting material phenyl ethyne is shown [$p(H_2) = 15$ bar, $p(CO_2) = 150$ bar, $T = 50\,°C$]

lent transfer of the product from the polar to the nonpolar phase. The slow phase transfer often leads to long reaction times even if the phases are vigorously mixed by stirring. As a consequence, successive side reactions or isomerizations are promoted, which increase the costs of downstream processing such as product or enantiomer separation.

4.2.1.1 Single-Phase Colloid Catalysis

In contrast, the extraordinary efficiency of a true single-phase SCF colloid catalysis was recently demonstrated for various alkynes that were hydrogenated with bimetallic colloids in $scCO_2$ [44]. The colloids consisted of gold (core) and palladium (shell) as obtained by the simultaneous reduction of the corresponding metal salts $AuCl_4$ and $Pd(Ac)_2$ in toluene. The colloids' particle size was reasonably narrow-disperse with an average diameter of 2 nm [162]. Before initiating the reduction of the salts, a sufficient amount of a block copolymer ligand consisting of styrene and 4-vinylpyridine ($M_n = 22\,500$) was added to the toluene solution for stabilizing the colloids after their formation. About 5–10 catalytically active colloids are embedded in each polymer ligand, yielding inverted polymer micelles with an average diameter of about 12 nm [163, 164]. Besides palladium and gold, other bimetallic combinations were used to synthesize colloid catalysts. Prominent examples are palladium and platinum or palladium and zinc. The colloids' electronic structure, however, and their surface composition are substantially influenced by the choice of metals, and catalytic activities may change significantly.

Catalytic activities in $scCO_2$ and the overall kinetics of catalytic reactions are most conveniently and meaningfully monitored by in situ NMR spectroscopy in TCAs (cf. Sect. 3.1). Accordingly, the single-phase colloidal hydrogenation of alkynes was also investigated in a TCA probe. After charging the toroid pressure vessel outside of the NMR magnet with catalyst and reagent, sealing it pressure-tight, and inserting the probe with the reactive solution into the magnet, the TCA was pressurized with CO_2 and heated to reach supercritical conditions. Subsequent to the dissolution of catalyst and reagent, the NMR magnetic field was shimmed for optimum homogeneity, and gaseous H_2 was injected through the second pressure capillary with the nonreturn valve (Fig. 10). Independent experiments conducted earlier with the same TCA verified that, at pressures above the p_c of CO_2, the pressure increase because of H_2 addition is almost identical to the partial pressure of H_2.

Immediately after the injection of H_2, the acquisition of multiple NMR spectra was started. Typically, 20 spectra were acquired with a delay of 20 s between two spectra. Integrated signal intensities were converted into absolute amounts by scaling them to an internal standard. Figure 18, for example, shows the time-dependent change of the amount of each component in the colloid-catalyzed hydrogenation of phenyl ethyne to styrene and, subsequently, from styrene to ethyl benzene. All data of Fig. 18 were acquired with the same reactive solution but without interrupting or "quenching" the reaction progress to take samples for external analysis. The catalytic reaction follows precisely the typical so-called A–B–C kinetics, in which a reagent (A) forms an intermediate product (B) that, subsequently, reacts to yield the final product (C). Surprisingly and in contrast with the early findings of Zemaian et al. [148], the single-phase colloidal hydrogenations of acetylenes in $scCO_2$ approach completion after only a few minutes.

4.2.1.2 Turnover Frequency and Catalytically Active Site

The efficiency of a catalytic reaction and the reactivity of a catalyst are commonly described by turnover numbers (TONs) and TOFs [165]. While the TON represents the amount of material that is converted in a particular reaction by a single catalytically active site, the TOF is defined as the amount of material that is converted by an active site in a given time. The latter is usually calculated from the amount of product formed during a certain time over the number of catalytically active sites. Its common unit is per hour. In standard homogeneous transition-metal catalysis, the number of active sites corresponds with the number of transition-metal atoms in the solution. If a catalytic reaction uniformly generates product without acceleration or inhibition because of induction periods or changing concentrations, respectively, the TOF is equal to TON/t, where t is the time. However, a catalytic reaction rarely has such a constant conversion rate exhibiting zeroth-order kinetics. Therefore, the most commonly found—and most conservative—approach to determine TOFs is to use the time during which 50% of the starting material was converted to product. Frequently, however, other methods are used for determining TOFs, such as the instantaneous reaction rate at the beginning of the catalysis (i.e., the starting kinetics), which inevitably leads to values that are incomparable with those from the common approach.

In colloid catalysis, it is important to determine the number of active sites, because a single colloid is able to convert more than just one reagent molecule at the same time. Then again, using the amount of transition metal that is used to generate the colloids is unrealistic, because not all atoms might be exposed to the surface and, even if the were, not all transition-metal atoms are able to convert reagent molecules at the same time because of steric interference. A similar problem occurs in heterogeneous catalysis, where conversion rates are usually determined with respect to the surface area of the catalyst. However, it is questionable if the surface area of colloids plays a reliable role for comparing TOF, especially, since colloid surfaces are highly curved.

In a recent paper, a method for calculating TOFs in colloid catalysis was derived based on a straightforward model that relates an active site to the area of steric interference [44]. Accordingly, a catalytically active site consists of the area that is needed by a reagent molecule to freely approach the colloid. For the previously mentioned colloid catalysis in $scCO_2$, a conservative supposition led to an area of seven Pd atoms to generate an active site. Accordingly, with the additional assumption that all Pd atoms are on the surface of the colloids [164], the number of active sites was calculated from one seventh of the nominal amount of Pd used for the synthesis of the colloids.

4.2.1.3
Kinetics

In Fig. 19, TOFs determined from the time of 50% conversion of the starting material are displayed for colloidal hydrogenations of various alkynes at a hydrogen pressure of 15 bar and 50 °C. An SCR of about 6.5×10^3 was used for all reactions. From the extraordinarily high TOFs that are achieved with rather mild reaction conditions, it is evident that the colloid catalysis in $scCO_2$ exemplified here proceeds faster than any other known homogeneously catalyzed reaction. In addition, the reactions utilize the tunable qualities of both $scCO_2$ and colloid catalysts, and allow for effortless quantitative separation of the reaction products from the catalyst and solvent.

The influence of externally adjustable reaction conditions on the colloid catalysis in $scCO_2$ is shown in Fig. 20, in which TOFs for the reagents phenyl ethyne and 3-methyl 1-pentyne-3-ol are plotted as a function of H_2 pressure [$p(H_2) = 2–15$ bar]. Up to $p(H_2) = 15$ bar, the conversion rates increase

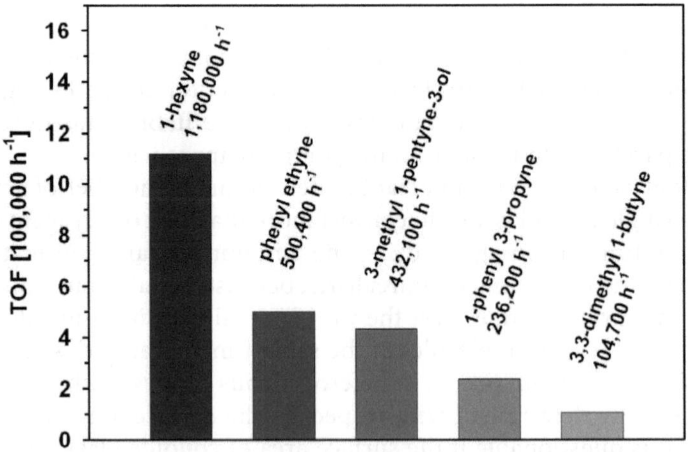

Fig. 19 Turnover frequencies (*TOFs*) for the colloid-catalyzed hydrogenation of various alkynes [44]. The reaction conditions are $p(H_2) = 15$ bar, $p(CO_2) = 150$ bar, and $T = 50$ °C

dramatically, indicating that the availability of H_2 is still a limiting factor for the reaction. Further increases of $p(H_2)$, however, do not raise the TOFs much further, showing that the reaction mixture is now sufficiently saturated with H_2.

Further increases in the TOF, however, were still reached by increasing the SCR for the colloid catalysis in $scCO_2$. Figure 21 shows the consumption of 1-hexyne (i.e., the reagent for which the highest TOFs have been measured before, cf. Fig. 18) at SCRs of 6.6×10^3, 13.2×10^3, and 19.8×10^3. It is noted that both axes of the graph in Fig. 21 are normalized for ease of comparison between the three curves. The TOFs rise from about 10^6 h^{-1} (SCR = 6.6×10^3) to over 4×10^6 h^{-1} (SCR = 19.8×10^3). In addition, the characteristics of the curves change, which indicates a changing reaction order. At an SCR of 6.6×10^3, an exponential decay prevails, representing first-order kinetics with respect to the reagent. At SCRs of 13.2×10^3 and 19.8×10^3, however, a linear decay is observed at the beginning of the reaction, which indicates zeroth-order kinetics and, thus, the independence of the catalytic reaction from H_2 and reagent concentrations. Zeroth-order kinetics is rarely observed with catalytic hydrogenations in conventional solvents, reflecting rate limitation of the catalysis from molecular diffusion, poor solubility of H_2, and unfavorable kinetics of the H_2 phase transfer between the gaseous and the liquid phase.

A similar approach to enhance the catalytic reactivity was described by Crooks et al. [166] utilizing dendrimer-encapsulated palladium nanoparticles for the catalytic hydrogenation of olefins in liquid CO_2 and in $scCO_2$.

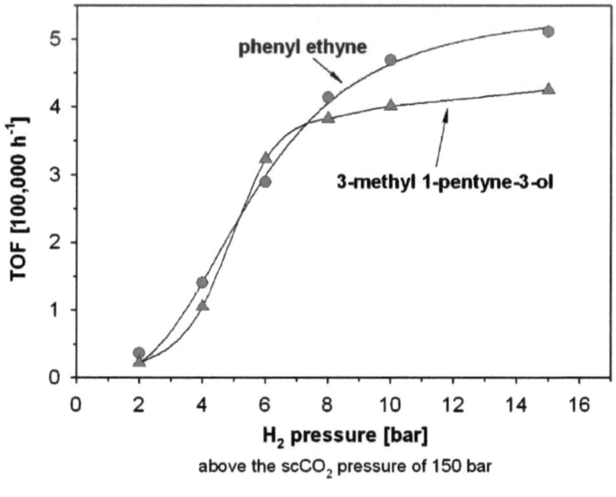

Fig. 20 TOFs for the colloid-catalyzed hydrogenation of phenyl ethyne (*circles*) and 3-methyl 1-pentyne-3-ol (*triangles*) as a function of H_2 pressure [44]. Further reaction conditions are $p(CO_2) = 150$ bar and $T = 50$ °C

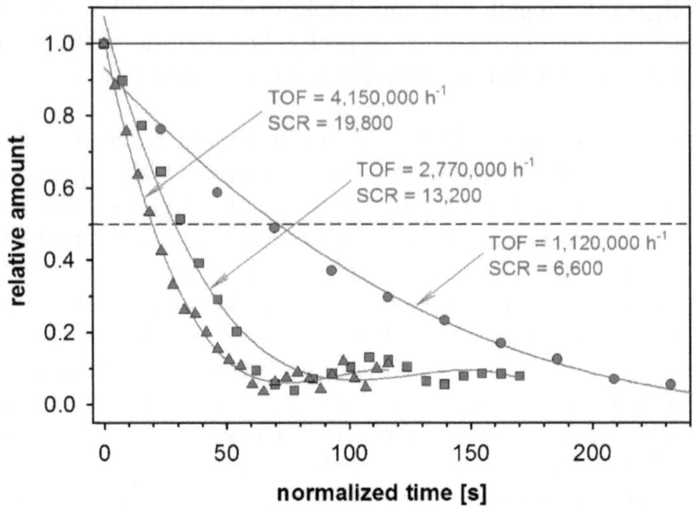

Fig. 21 Relative amount of 1-hexyne consumed in the colloid-catalyzed hydrogenation as a function of reaction time [44]. The substrate-to-catalyst ratios (*SCRs*) are 6600 (*circles*), 13 200 (*squares*), and 19 800 (*triangles*), and the reaction conditions are $p(H_2) = 15$ bar, $p(CO_2) = 150$ bar, and $T = 50\,°C$. Abscissa and ordinate are both normalized to assist the comparison between the three curves

The nanoparticles consisted of catalytically active Pd(0) colloids that were stabilized in perfluoropolyether-functionalized poly(propylene imine) dendrimers. The fluorination was necessary to achieve reasonable solubility in $scCO_2$. A high efficiency of the fluorinated dendrimer-encapsulated catalyst was illustrated by catalytic hydrogenation of olefins such as styrene to form ethyl benzene. In addition, the colloid catalyst was used for a Heck coupling of iodobenzene with methyl acrylate to form exclusively the desired product methyl 2-phenyl acrylate.

5 Interactions of Carbon Dioxide with Fluorinated Compounds

As exemplified several times in the previous sections, many compounds—whether they be substrates or catalysts—show increased affinity to CO_2 if they are modified by fluorination [167–169]. Therefore, the addition of fluorinated pony tails to existing chemical structures or ligand frameworks as exemplified in Sect. 4.2 is widely accepted as a valid strategy to improve the solubility in $scCO_2$ [170]. Additionally, $scCO_2$ is the only simple solvent known for swelling poly(tetrafluoroethylene) (Teflon) or dissolving perfluorinated dendrimers up to, at least, generation 4. In contrast, fluorinated compounds are often poorly soluble in conventional solvents.

5.1 Theoretical Approaches

From the findings described, it is astounding that no consensus exists about the reasons for such a preferred dissolution of fluorinated compounds in $scCO_2$. Instead, two theoretical approaches are discussed rather controversially. One theory suggests that fluorination decreases intermolecular interaction between molecules, leading to a favorable formation of mixtures [171]. In addition, the vapor pressure of fluorinated compounds is usually reduced compared with that of the equivalent nonfluorinated compounds, making them more volatile and miscible with $scCO_2$. It is recalled in this context that SCFs are miscible with gases or other highly volatile compounds at almost all ratios. Following this theory, the increased solubility of fluorinated compounds is simply an entropy effect, and active interactions between compound and $scCO_2$ are excluded or are only minimal. Another theory, however, states that the increased solubility of fluorinated molecules in $scCO_2$ is caused by active interactions between fluorinated moieties and CO_2 molecules, which are not achieved with nonfluorinated compounds [172]. In this case, the increased solubility is not only based on entropy but also on enthalpy effects.

5.2 NMR Experimental Validations

To support the latter theory, in situ NMR spectra of partially fluorinated acrylic acid ester (Fig. 22) dissolved in $scCO_2$ were recorded from a TCA probe.

As expected, the model compound acrylic acid ester exhibits a very high solubility in its fluorinated form but is only marginally soluble in $scCO_2$ without the fluorination. ^{19}F NMR spectra of the fluorinated ester in $scCO_2$ exhibit the signals labeled 3-CF_2, 9-CF_2, and 10-CF_3 in Fig. 23, which are unambiguously assigned to the equivalently labeled CF_2 and CF_3 groups of the molecule shown schematically in Fig. 22. The signals of the intermediate CF_2 groups (4-CF_2 to 8-CF_2) were not assigned individually but rather treated as a group of signals (4/8-CF_2) representing moieties of similar properties.

In NMR spectroscopy, the relative resonance frequency (i.e., chemical shift measured in parts per million) is an important parameter to evaluate electronic effects such as intermolecular and intramolecular interactions. In Fig. 24, the chemical shifts of the fluorinated groups 3-CF_2, 4/8-CF_2, 9-CF_2, and 10-CF_3 are plotted as a function of temperature. In this plot, all chemical shifts were set to the nominal value of zero at 25 °C, i.e., at a tem-

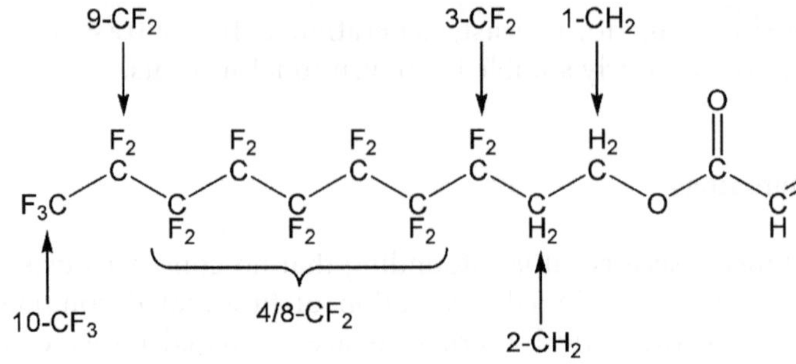

Fig. 22 Schematic structure of the partially fluorinated reagent (3,3,4,4,5,5,6,6,7,7,8,8,9,9,10,10,10-heptadecafluorodecyl) acrylic acid ester. The *numbers* in the labels (e.g., 3-CF_2) indicate the position of the CH_2, CF_2, or CF_3 moieties with respect to the ester functionality

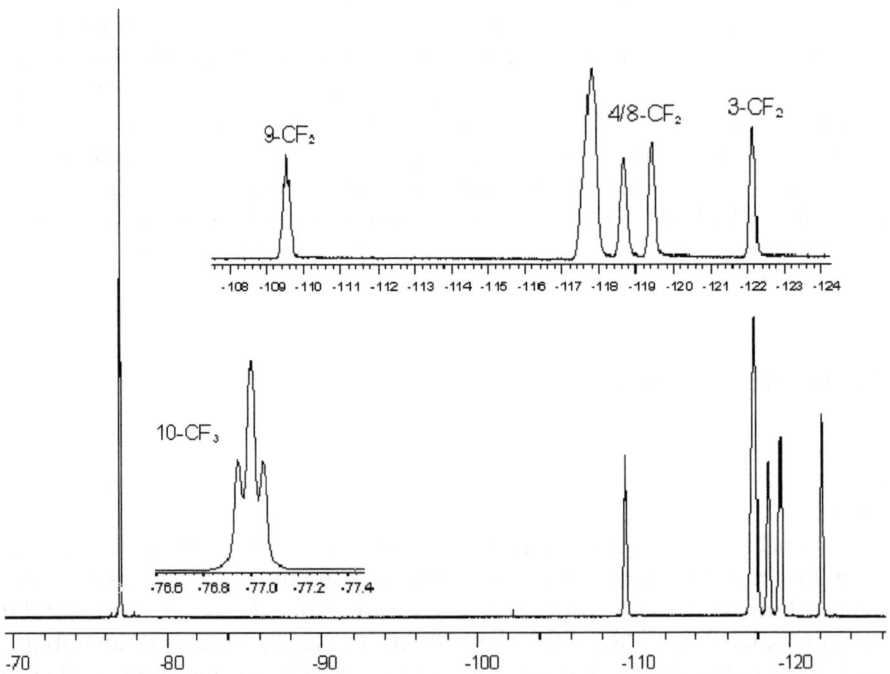

Fig. 23 ^{19}F NMR spectra of the fluorinated acrylic acid ester (Fig. 22) in scCO_2. The assignments of the groups 3-CF_2, 9-CF_2, 4/8-CF_2, and 10-CF_3 correspond to the labels in Fig. 22. Intermediate CF_2 groups (4-CF_2 to 8-CF_2) are not assigned individually but are treated as a group with similar properties. They are labeled 4/8-CF_2

perature where CO_2 is still a conventional liquid under the applied pressure of $p(CO_2) = 100\,\text{bar}$. Generally, changes of chemical shifts as a function of temperature are not unusual for molecules in solution even if no additional interactions are expected. For example, the aliphatic electronic shielding for

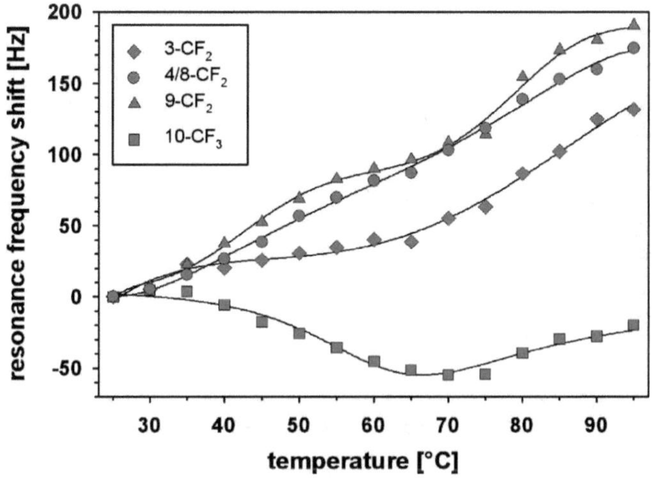

Fig. 24 Change of chemical shift for the fluorinated groups 3-CF_2, 4/8-CF_2, 9-CF_2, and 10-CF_3 of the fluorinated acrylic acid ester (Fig. 22) in scCO_2 as a function of temperature [$p(CO_2) = 100$ bar]. To easily compare the changes, all chemical-shift values were arbitrarily set to zero at 25 °C

NMR-active nuclei decreases with rising temperatures, because excited vibration energy levels become more heavily populated, which increases the average bond length and, in turn, reduces the electron density at the bonding atoms. Accordingly, a frequency change towards higher chemical-shift values is expected as observed in Fig. 24 for all CF_2 groups. In contrast, however, the chemical shift of 10-CF_3 initially (30–70 °C) moves in the opposite direction before it follows the general trend and changes towards higher values. This special behavior indicates an increasing electron density for the fluorine atoms of the 10-CF_3 group, which is most likely the result of a specific interaction with free electron pairs of CO_2 as it becomes supercritical. However, no such interaction is detected for the other fluorine atoms, i.e., for the fluorine atoms of the intermediate 3-CF_2, 4/8-CF_2, or 9-CF_2 groups.

Not only chemical-shift values but also nuclear spin relaxations are strongly influenced by intermolecular and intramolecular interactions. Accordingly, ^{19}F spin-lattice relaxation times (T_1) that reflect the mobility of the fluorine atoms were measured by a standard inversion-recovery procedure in scCO_2 as a function of temperature (Fig. 25). Just akin to the chemical-shift values, the 10-CF_3 group exhibits a substantially different temperature-dependent behavior compared with the rest of the CF_2 groups. While the interpretation of temperature-dependent T_1 changes is less straightforward, the different behavior of the 10-CF_3 group in scCO_2 is clearly visible above 40 °C as the system changes from a liquid to an SCF. Again, a special interaction between 10-CF_3 is concluded. Alternatively, however, the formation of micelles was suggested, in which the outermost CF_3 group is still very mobile

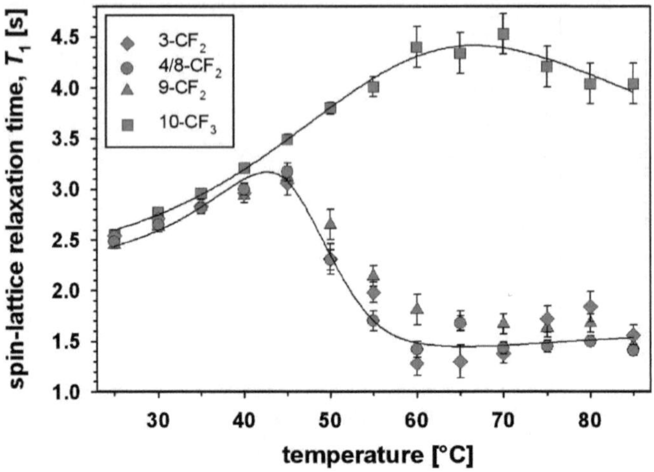

Fig. 25 ^{19}F spin-lattice relaxation times (T_1) for the fluorinated groups 3-CF_2, 4/8-CF_2, 9-CF_2, and 10-CF_3 of the fluorinated acrylic acid ester (Fig. 22) in $scCO_2$ as a function of temperature [$p(CO_2) = 100$ bar]

but the CF_2 groups are confined into a more rigid, membranelike structure. The theory of micelle formation would particularly explain why the T_1 values of the CF_2 groups change dramatically while the 10-CF_3 group follows a more uniform trend.

$ScCO_2$ exhibits a dissolution power similar to that of cyclohexane if it is adjusted to about the same density. Accordingly, a comparison of the ^{19}F NMR parameters (chemical shift and T_1 relaxation time) in $scCO_2$ and in cylcohexane was conducted for the fluorinated model compound. Figure 26 shows the T_1 relaxation times of the CF_2 and CF_3 groups of the fluorinated acrylic acid ester in cyclohexane measured by inversion recovery as a function of temperature. In contrast with the solvent $scCO_2$, the changes of T_1 times in cyclohexane are not spectacular but follow a uniform trend for all groups including 10-CF_3. The T_1 values for the CF_3 group show slightly oscillating behavior, which, however, might be due to temperature-dependent activations of different rotation modes for this group. The mapping of the chemical shifts of the CF_2 and CF_3 groups as a function of temperature reveals a similar unspectacular trend and does not indicate the development of special interactions between cylcohexane and the fluorinated groups of the acrylic acid ester.

From the previous results, we conclude that specific interactions between $scCO_2$ and, at least, the CF_3 groups of fluorinated alkyl chains are most likely. A similar interaction between $scCO_2$ and intermediate CF_2 groups, however, could not be determined. Thermodynamically, the specific interactions revealed are enthalpy effects, which supports the second of the previously mentioned theories for preferred dissolution of fluorinated compounds in $scCO_2$.

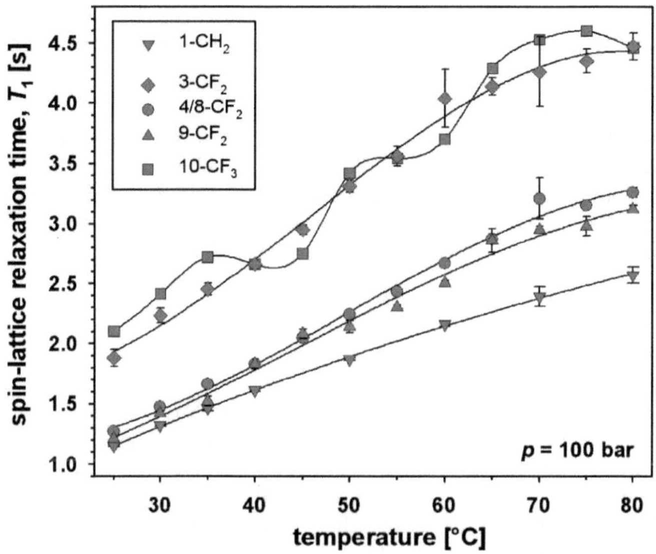

Fig. 26 ^{19}F spin-lattice relaxation times (T_1) for the fluorinated groups 3-CF_2, 4/8-CF_2, 9-CF_2, and 10-CF_3 of the fluorinated acrylic acid ester (Fig. 22) in cyclohexane as a function of temperature. For comparison, ^{1}H spin-lattice relaxation times of 1-CH_2 are included

Entropy effects, however, may also play a significant role in this context for it is known that long perfluorinated alkyl chains promote the dissolution in $scCO_2$ more than short chains. Because no specific interaction was found for intermediate CF_2 groups, this promotion might be due to entropy effects as stated by the first theory.

6 Concluding Remarks

Research in the area of SCFs and, especially, with $scCO_2$ has boomed substantially during the last decade. Not all of the research within this field is motivated by the environmentally benign properties of a modern solvent, and not all activities are aimed towards substitution of traditional, rather toxic organic solvents with the so-called green $scCO_2$. Especially, recent developments in colloid catalysis have shown that $scCO_2$ bears the qualities to establish its own chemistry leading to new pathways in synthesis that cannot be followed with traditional methods or solvents. To promote this kind of research, a broad and new infrastructure of analytical tools is necessary as exemplified in this article by the in situ NMR spectroscopy in modern autoclave probes and by the specialized technique of PHIP spectroscopy. These tools will help to understand fluid properties, reaction mechanisms, and kinetics in $scCO_2$ more thoroughly and, in turn, further encourage and spark research activities with SCFs.

Acknowledgements The work of the authors in connection with the research topic of this article was supported by the Deutsche Forschungsgemeinschaft (DFG), the Max Buchner Research Foundation (MBFSt) of the German Society for Chemical Engineering and Biotechnology (DECHEMA), and through a graduate fellowship of the German National Academic Foundation (Studienstiftung des Deutschen Volkes).

References

1. Wells SL, DeSimone JM (2001) Angew Chem Int Ed Engl 40:518
2. Taylor DK, Carbonell R, DeSimone JM (2000) Annu Rev Energy Environ 25:115
3. Hitzler MG, Smail FR, Ross SK, Poliakoff M (1998) Org Proc Res Dev 2:137
4. Subramaniam B, Lyon CJ, Arunajatesan V (2002) Appl Catal B
5. Webb PB, Cole-Hamilton DJ (2004) Chem Commun 612
6. Zosel K (1978) Angew Chem 90:748
7. Wilke G (1978) Angew Chem 90:747
8. Schneider GM (1978) Angew Chem 90:762
9. Hubert P, Vitzthum OG (1978) Angew Chem 90:756
10. Eggers R (1978) Angew Chem 90:799
11. Gnayfeed MH, Daood HG, Illes V, Biacs PA (2001) J Agric Food Chem 49:2761
12. Eller FJ, King JW (2001) J Agric Food Chem 49:4609
13. Manninen P, Pakarinen J, Kallio H (2001) J Agric Food Chem 45:2533
14. McHugh M, Krukonis V (1986) Supercritical fluid extraction— principles and practice. Butterworth, Boston
15. Noyori R, Ikariya T (2000) In: Vöglte F, Stoddart JF, Shibasaki M (eds) Stimulating concepts in chemistry. Wiley-VCH, Weinheim, p 13
16. Li J, Beckman EJ (1998) Ind Eng Chem Res 37:4768
17. Samsonov MD, Wai CM, Lee S, Kulyako Y, Smart NG (2001) Chem Commun 1868
18. Drescher M, Bonakdar M (2001) Chem Ing Tech 73:338
19. Dahmen N, Schön J, Dinjus E (2000) Chem Ing Tech 72:950
20. Qi, XH, Zhuang YY, Yuan YC, Gu WX (2002) J Hazard Mater 90:51
21. Jessop PG, Ikariya T, Noyori R (1999) Chem Rev 99:475
22. Baiker A (1999) Chem Rev 99:453
23. Kendall JL, Canelas DA, Young JL, DeSimone JM (1999) Chem Rev 99:543
24. Mesiono AJ, Beckman EJ, Russel AJ (1999) Chem Rev 99:623
25. Hancu D, Beckman EJ (1999) Ind Eng Chem Res 38:2824
26. Jessop PG, Leitner W (eds) (1999) Chemical synthesis using supercritical fluids. Wiley-VCH, Weinheim
27. Oakes RS, Clifford AA, Rayner CM (2001) J Chem Soc Perkin Trans 1:917
28. Aresta M, Quaranta E (1997) Chem Tech 27:32
29. Fujita S, Bhanage BM, Hikushima Y, Arai, M (2001) Green Chem 3:87
30. McGhee WD, Riley DP, Christ ME, Christ KM (1993) Organometallics 12:1429
31. McGhee WD, Pan Y, Riley DP (1994) J Chem Soc Chem Commun 699
32. Dinjus E, Leitner W (1995) Appl Organomet Chem 9:43

33. Kubota T, Hayakawa I, Mabuse H, Mori K, Ushikoshi K, Watanabe T, Saito M (2001) Appl Organomet Chem 15:121
34. French SA, Sokol AA, Bromley ST, Catlow CRA, Rogers SC, King F, Sherwood P (2001) Angew Chem 113:4569
35. Watanabe M, Osada M, Inomata H, Arai K, Kruse A (2003) Appl Catal A 245:333
36. Reubroycharoen P, Vitidsant T, Asami K, Yoneyama Y, Tsubaki N (2003) Catal Commun 4:461
37. Thomas CA, Bonilla RJ, Yong H, Jessop PG (2001) Can J Chem 79:719
38. Burgmeister T, Kastner F, Leitner W (1993) Angew Chem Int Ed Engl 32:739
39. Jessop PG, Hsiao Y, Ikariya T, Noyori R (1994) J Am Chem Soc 116:8851
40. Cooper AI (2000) J Mater Chem 10:207
41. deVries TJ, Duchateau R, Vortsman MAG, Keurentjes JTF (2000) Chem Commun 263
42. Buback M, Jürgens M (2003) Ind Eng Chem Res 43:6338
43. Landolt-Börnstein (1996) Transportphänomene 1, 6th edn,vol 2. Springer, Berlin Heidelberg New York
44. Niessen HG, Eichhorn A, Woelk K, Bargon J (2002) J Mol Catal A 182-183:463
45. Savage PE, Yu J, Stylski N, Brock EE (1998) Supercrit Fluids 12:141
46. Hoffmann MM, Conradi MS (1997) Rev Sci Instrum 68:159
47. Zhang X, Savage PE (1998) Catal Today 40:333
48. Savage PE (1999) Chem Rev 99:603
49. Trautner P, Woelk K (2002) Phys Chem Chem Phys 4:5973
50. Etesse P, Zega JA, Kobayashi R (1992) J Phys Chem 97:2022
51. Vardag T, Kargere N, Lüdemann HD (1991) Ber Bunsenges Phys Chem 95:859
52. Levine IN (1988) Physical chemistry, 3rd edn. McGraw-Hill, New York
53. Bruch MD (ed) (1996) NMR spectroscopy techniques. Practical spectroscopy series, vol 21. Dekker, New York
54. Rathke JW, Klingler RJ, Chen MJ, Woelk K (1994) Trends Organomet Chem 1:117
55. Gaemers S (1999) PhD thesis. University of Amsterdam
56. Trautner P (2001) PhD thesis. University of Bonn
57. Niessen HG (2002) PhD thesis. University of Bonn
58. Roe DC, Kating PM, Krusic PJ, Smart BE (1998) Top Catal 5:133
59. Landis CR, Halpern J (1987) J Am Chem Soc 109:1746
60. Woelk K, Bargon J (1993) Z Phys Chem 182:155
61. Vander Velde D, Jonas J (1987) J Magn Reson 71:480
62. Hovarth IT, Millar JM (1991) Chem Rev 91:1339
63. Woelk K, Bargon J (1992) Rev Sci Instrum 63:3307
64. Roe DC (1992) Adv Chem Ser 230:33
65. Elsevier CJ (1994) J Mol Catal 92:258
66. Fremgen DE, Smotkin ES, Gerald RE, Klingler RJ, Rathke JW (2001) J Supercrit Fluids 19:287
67. Rathke JW, Klingler RJ, Chen MJ, Woelk K (1997) Prog NMR Spectrosc 30:209
68. Woelk K, Rathke JW, Klingler RJ (1994) J Magn Reson A 109:137
69. Woelk K, Rathke JW, Klingler RJ (1993) J Magn Reson A 105:113
70. Barrall GA, Lee YK, Chingas GC (1994) J Magn Reson A 106:132
71. Momot KI, Binesh N, Kohlmann O, Johnson CS (2000) J Magn Reson 142:348
72. Woelk K, Zwank BLJ, Trautner P, Lehnhof E, Bargon J, Klingler RJ, Gerald RE, Rathke JW (2000) J Magn Reson 145:276
73. Trautner P, Woelk K (2002) Phys Chem Chem Phys 4:5973

74. Gerald II RE, Klingler RJ, Rathke JW, Sandí G, Woelk K (1998) In: Blümler P, Blümich B, Botto R, Fukushima E (eds) Spatially resolved magnetic resonance. Wiley-VCH, Weinheim, p 111
75. Trautner P, Woelk K (2002) Appl Magn Reson 22:291
76. Niessen HG, Trautner P, Wiemann S, Bargon J, Woelk K (2002) Rev Sci Instrum 73:1259
77. Doty FD, Entzminger G, Yang YA (1998) Concepts Magn Reson 10:133
78. Niessen HG, Trautner P, Backkausen R, Woelk K (2003) Concepts Magn Reson B Magn Reson Eng 16:15
79. Bowers CR, Jones DH, Kurur ND, Labinger JA, Pravica MG, Weitekamp DP (1990) Adv Magn Reson 14:269
80. Eisenberg R (1991) Acc Chem Res 24:110
81. Eisenberg R, Eisenschmid TC, Chinn MS, Kirss RU (1992) Adv Chem Ser 230:47
82. Bargon J, Kandels J, Woelk K (1993) Z Phys Chem 180:65
83. Bargon J (1996) In: Cornils B, Herrmann WA (eds) Applied homogeneous catalysis with organometallic compounds, vol 2. VCH, Weinheim, p 672
84. Natterer J, Bargon J (1997) Prog NMR Spectrosc 31:293
85. Duckett SB, Sleigh CJ (1999) Prog NMR Spectrosc 34:71
86. Duckett SB, Colebrooke SA (2002) Enc Nuc Magn Reson 9:598
87. Bargon J, Fligg R (2000) Research matters. Oxford Instruments Superconductivity, issue 12, p 7
88. Bargon J, Kandels J, Woelk K (1990) Angew Chem Int Ed Engl 29:58
89. Jonischkeit T, Woelk K (2004) Adv Synth Catal 346:875
90. Baiker A (1999) Chem Rev 99:453
91. Tadd AR, Marteel A, Mason MR, Davies JA, Abraham MA (2002) Ind Eng Chem Res 41:4514
92. Wandeler R, Künzle N, Schneider MS, Mallat T, Baiker A (2001) Chem Commun 673
93. Kamat SV, Beckmann EJ, Russel AJ (1995) Crit Rev Biotechnol 15:41
94. Zagrobelny J, Bright FV (1992) Biotechnol Prog 8:421
95. Matsuda T, Harada T, Nakamura K (2000) Chem Commun 1367
96. Rathke JW, Klingler RJ, Krause TR (1991) Organometallics 10:1350
97. Guo Y, Akgerman A (1997) Ind Eng Chem Res 36:4581
98. Guo Y, Akgerman A (1999) J Supercrit Fluids 15:63
99. Jessop PG, Ikariya T, Noyori R (1995) Organometallics 14:1510
100. Beller M, Cornils B, Fohning CD, Kohlpaintner CW (1995) J Mol Catal 104:17
101. Cornils B, Herrmann WA (eds) (1996) Applied homogeneous catalysis with organometallic compounds. VCH, Weinheim
102. Arakawa H, Aresta M, Amor JN, Barteau MA, Beckman EJ, Bell AT, Bercaw JE, Creutz C, Dinjus E, Dixon DA, Domen K, DuBois DL, Eckert J, Fujita E, Gibson DH, Goddard WA, Goodman DW, Keller J, Kubas GJ, Kung HH, Lyons JE, Manzer LE, Marks TJ, Mrokuma K, Nichols KM, Periana R, Que L, Rostrup-Nielson J, Sachtler WMH, Schmidt LD, Sen A, Somorjai GA, Stair PC, Tumas W (2001) Chem Rev 101:953
103. Herrmann WA, Kohlpaintner CW (1993) Angew Chem 105:1588
104. Rathke JW, Klingler RJ, Krause TR (1992) Organometallics 11:585
105. Klingler RJ, Rathke JW (1994) J Am Chem Soc 116:4772
106. Ke J, Han B, George MW, Yan H, Poliakoff M (2001) J Am Chem Soc 123:3661
107. Koch D, Leitner W (1998) J Am Chem Soc 120:13398
108. Kainz S, Leitner W (1998) Catal Lett 55:223
109. Bach I, Cole-Hamilton DJ (1998) Chem Commun 1463

110. Hu Y, Chen W, Banet Osuna AM, Iggo JA, Xiao JL (2002) Chem Commun 788
111. Sellin MF, Webb PB, Cole-Hamilton DJ (2001) Chem Commun 781
112. Webb PB, Sellin MF, Kunene TE, Williamson S, Slawin AMZ, Cole-Hamilton DJ (2003) J Am Chem Soc 125:15577
113. Stemmer H (2001) PhD thesis. University of Jena
114. Erkey C, Lozano Diz E, Süss-Fink G, Dong X (2002) Catal Commun 3:213
115. Haas GR, Kolis JW (1995) Organometallics 17:4454
116. Natterer J (1997) PhD thesis. University of Bonn
117. Permin AB, Eisenberg R (2002) J Am Chem Soc 124:12406
118. Jessop PG, Ikariya T, Noyori R (1994) Nature 368:231
119. Jessop PG, Hsiao Y, Ikariya T, Noyori R (1996) J Am Chem Soc 118:344
120. Niessen HG, Ulrich C, Bargon J (2000) J Labelled Compd Radiopharm 43:711
121. Sanchez-Delgado RA, Bradley JS, Wilkinson G (1976) J Chem Soc Dalton Trans 399
122. Baiker (2000) Appl Organometal Chem 14:751
123. Kainz S, Koch D, Baumann W, Leitner W (1997) Angew Chem Int Ed Engl 36:1628
124. Leitner W (2000) Appl Organometal Chem 14:809
125. Lange S, Brinkmann A, Trautner P, Woelk K, Bargon J, Leitner W (2000) Chirality 12:450
126. Palo DR, Erkey C (1998) Ind Eng Chem Res 37:4203
127. Palo DR, Erkey C (2000) Organometallics 2000 19:81
128. Palo DR, Erkey C (1999) Ind Eng Chem Res 38:3786
129. Hu Y, Chen W, Xu L, Xiao J (2001) Organometallics 20:3206
130. Guzel B, Omary MA, Fackler JP, Akgerman A (2001) Inorg Chim Acta 325:45
131. Lin B, Akgerman A (2001) Ind Eng Chem Res 40:1113
132. Burk MJ, Feng S, Gross MF, Tumas W (1995) J Am Chem Soc 117:8277
133. Xioa JL, Nefkens SCA, Jessop PG, Ikariya, Noyori R (1996) Tetrahedron Lett 37:2813
134. Hu Y, Chen W, Banet Osuna AM, Stuart AM, Hope EG, Xioa JL (2001) Chem Commun 725
135. Bart M, Küther J, Seshadri R, Tremel W (1998) Angew Chem 110:2646
136. Bönnemann H, Richards RM (2001) Eur J Inorg Chem 2455
137. Bönneman H, Braun G, Brijoux W, Brinkmann W, Schulze Tilling A, Seevogel K, Siepen K (1996) J Organomet Chem 520:
138. Schmidt G (1992) Chem Rev 92:1709
139. Tu W, Liu H, Liew KY (2000) J Colloid Interface Sci 229:453
140. Toshima K, Takahashi T (1992) Bull Chem Soc Jpn 65:4000
141. Crooks RM, Zhao M, Sun L, Checik V, Yeung LK (2001) Chem Res 34:181
142. Aiken JD, Finke RG (1999) J Am Chem Soc 121:8803
143. Lopez-Castillo ZK, Flores R, Kani I, Fackler JP, Akgerman A (2002) Ind Eng Chem Res 41:3075
144. Lopez-Castillo ZK, Flores R, Kani I, Fackler JP, Akgerman A (2003) Ind Eng Chem Res 42:3893
145. Goetheer ELV, Verkerk AW, van den Broeke LJP, de Wolf E, Deelman BJ, van Koten G, Keurentjes JTF (2003) J Catal 219:126
146. de Smet K, Aerts, S, Ceulemans E, Vankelecom IFJ, Jacobs PA (2001) Chem Commun 597
147. van den Broeke, LJP, Goetheer ELV, Verkerk AW, de Wolf E, Deelman BJ, van Koten G, Keurentjes JTF (2001) Angew Chem 113:4605
148. Zemanian TS, Bean RM, Fulton JL, Linehan JC, Smith RD (1991) 2nd international symposium on supercritical fluids, Boston, USA

149. Fremgen DE, Smotkin ES, Gerald RE, Klingler RJ, Rathke JW (2001) J Supercrit Fluids 19:278
150. Lee D, Hutchinson JC, DeSimone JM, Murray RW (2001) J Am Chem Soc 123:8406
151. Komoto I, Kobayashi S (2004) J Org Chem 69:680
152. Webb PB, Marr PC, Parson AJ, Gidda HS, Howdle SM (2000) Pure Appl Chem 72:1347
153. Bonillar RJ, James BR, Jessop PG (2000) Chem Commun 941
154. Heldebrant DJ, Jessop PG (2003) J Am Chem Soc 125:5600
155. Brown RA, Pollet P, McKoon E, Eckert CA, Liotta CL, Jessop PG (2001) J Am Chem Soc 123:1254
156. Liu F, Abrams MB, Baker RT, Tumas W (2001) Chem Commun 434
157. Wasserscheidt P, Keim W (2000) Angew Chem 112:3926
158. Lozana P, DeDiego T, Carri D, Vaultier M, Iborra JL (2002) Chem Commun 692
159. Sheldon R (2001) Chem Commun 2399
160. Bösmann A, Franci G, Janssen E, Solinas M, Leitner W, Wasserscheidt P (2001) Angew Chem 113:2769
161. Schulz J, Rocoux A, Patin H (1999) Chem Commun 535
162. Antonietti M, Heinz S, Schmidt M, Rosenauer C (1994) Macromolecules 27:3276
163. Yu H, Gibbons PC, Kelton KF, Buhro WE (2001) J Am Chem Soc 123:9198
164. Bronstein LM, Chernyshov DM, Volkow IO, Ezernitskaya MG, Valetsky PM, Matveev VG, Sulman EM (2000) J Catal 196:302
165. Gates BC (1992) Catalytic chemistry. Wiley, New York
166. Yeung, LK, Lee, T, Johnston KP, Crooks RM (2001) Chem Commun 2290
167. DeSomine JM, Guan Z, Elsbernd CS (1992) Science 257:945
168. Mertdogan CA, Byun HS, McHugh MA, Tuminello, WH (1996) Macromolecules 29:6548
169. McHugh MA, Park IH, Reisinger JJ, Ren Y, Lodge TP, Hillmyer MA (2002) Macromolecules 35:4653
170. Leitner W (1999) Top Curr Chem 206:107
171. Yee GG, Fulton JL, Smith RD (1992) J Phys Chem 96:6172
172. Dardin A, DeSimone JM, Samulski ET (1998) J Phys Chem 102:1775

Top Curr Chem (2007) 276: 111–124
DOI 10.1007/128_2006_096

Published online: 11 October 2006

Continuous Asymmetric Hydrogenation

Lasse Greiner[1] · Stephan Laue[2] · Jens Wöltinger[3] · Andreas Liese[4] (✉)

[1]RWTH Aachen University, Institut für Technische und Makromolekulare Chemie, Worringerweg 1, 52074 Aachen, Germany

[2]Bu-PLS-P & T-PD, Bayer Material Science, 51368 Uerdingen, Germany

[3]Degussa AG, Rodenbacher Chaussee 4, 63457 Hanau-Wolfgang, Germany

[4]Institute of Technical Biocatalysis, Hamburg University of Technology TUHH, Dennickestr. 15, 21073 Hamburg, Germany
liese@tuhh.de

Abstract Asymmetric hydrogenation with homogeneous soluble catalysts is a key technology for the production of enantiomerically enriched fine chemicals. Hydrogen supply via dense membranes in combination with continuous membrane filtration of soluble macromolecular catalysts allows for continuous hydrogenation. A optimised reactor with minimized volume was realized and used for continuous enzymatic and homogeneous catalysis. The vital importance of catalyst stability for continuous application is discussed. Furthermore, activation kinetics of the Pyrphos catalyst was investigated and found to be dependent on the presence of substrate.

1
Introduction

Reduction with molecular hydrogen is a clean, atom-efficient and widely applicable methodology for organic synthesis. Thermodynamically imposed limits for conversion do not practically exist, as the reaction energies for the reduction of organic double bonds are usually about $100\,\mathrm{kJ\,mol^{-1}}$. Furthermore, the reducing agent can be used in excess by application of higher pressures without affecting down stream processing. It can be removed easily as it is a permanent gas at ambient pressure and temperature. In heterogeneous catalysis these factors and vice versa the availability of a vast variety of hydrogen activating catalysts explain the variety and broad use for organic synthesis at the lab and multi-ton scale.

This was further extended by the discovery of Wilkinson and coworkers that molecular solved precious metal complexes such as the now famous

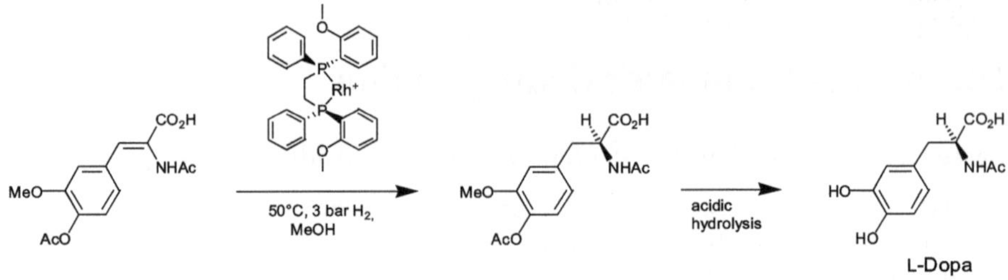

Fig. 1 Synthesis of L-dopa with Rh-DIPAMP catalyst (TTN >10 000, *ee* = 0.96)

triphenylphosphane rhodium catalyst are similarly capable of activating hydrogen and selectively adding hydrogen to double bonds [1]. This was then further extended by numerous efforts to yield enantioselective transformations. Knowles and Noyori were awarded the Noble Prize in Chemistry for their achievements in this field.

Knowles and Sabacky developed the first industrial asymmetric hydrogenation process starting from cinnamic acid derivatives to synthesize L-dopa (3,4-dihydroxyphenylalanine), which is used in the treatment of Parkinson's disease (Fig. 1) [2]. Their homogeneously soluble catalyst is a cationic rhodium bisphosphine complex, in which the enantioselectivity is induced by the chiral bisphosphine ligand. Total turnover numbers of the DIPAMP/Rh catalyst are > 10 000, and an *ee* = 0.96 is reached (DIPAMP: 1,2-bis-[(2-methoxy-phenyl)-methyl-phenyl-phosphanyl]-ethane). The optically pure L-dopa is separated from the catalyst and further purified by crystallization.

The work of Kagan and coworkers showed that the chirality could be moved from the phosphorus atom to the carbon backbone by introducing DIOP (2-dimethyl-4,5-bis-(diphenylphosphanyl)-pyrrolidine) as the ligand [3]. This general approach made access to ligands easier and facilitated the development of ligands for hydrogenation and asymmetric catalysis in general and this is ongoing today [4].

Brown [5, 6] and Halpern [7] revealed the complex network underlying the hydrogenation. They showed that two competing catalytic cycles lead to the formation of enantiomers and that the selectivity is controlled by the kinetics of intermediate steps. The interaction of the reactants is complex and unique for any combination of catalyst and substrate, partly explaining the vast variety of thousands of ligands found in the literature as slight variations of electronic and steric properties can have huge influences on activity and selectivity.

The most prominent ligands are derived from components of the "chiral pool", i.e., BPPM (ethyl 4-diphenylphosphanyl-2-[(diphenylphosphanyl)-methyl]-pyrrolidine-1-carboxylate) derived from praline [8], or Pyrphos or Deguphos (1-phenyl-3,4-bis-(diphenylphosphanyl)-pyrrolidine) derived from tartaric acid by the group of Nagel [9].

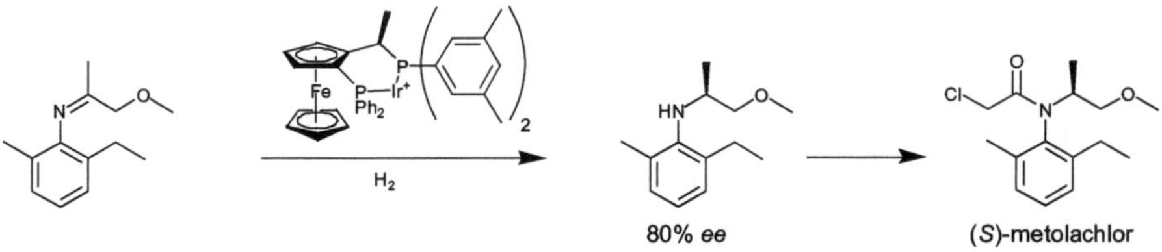

Fig. 2 Synthesis of (*S*)-metolachlor using the ferrocene-based XyliPhos/Ir catalyst

Recent developments diversify the ligand structural motifs further. In the ligand family represented by DuPhos (1,2-bis(2,5-dimethylphospholanyl)-benzene) the center of chirality is again located at the phosphorus incorporated in the five-membered rings. These very selective and active ligands were developed by Burk and coworkers [10]. Structural diversity is also given by examples like JosiPhos and PhanePhos representing structures with both axial as well as central chirality [11–13]. XyliPhos needs to be mentioned as it is used for the production of (*S*)-metolachlor on a scale of 10 000 t a^{-1} by Syngenta (Fig. 2). This ligand in combination with Ir as the metal center is unmatched for activity with TOF more than 300 min^{-1} and TTN of up to 10^6. However, enantioselectivity is with 80% rather low compared to other hydrogenation catalysts.

Asymmetric hydrogenation of ketones was made possible by the work of Noyori's group. They systematically used the axially chiral BINAP/Ru (BINAP: 2,2′-bis-phosphanyl-[1,1′]binaphthalenyl) as the catalyst for the enantioselective hydrogenations of prochiral α,β- and β,γ-unsaturated carboxylic acids, enamides, allylic and homoallylic alcohols, imines, etc [14]. BINAP is used for hydrogenations and a vast variety of other catalytic reactions. Industrially important are the hydrogenations leading to naproxen by Syntex (Fig. 3), and the (–)-menthol production by Takasago [15].

Homogeneously soluble catalysts are difficult to recycle or even to separate after use. This is documented by efforts made to try to generalize the down-stream separation and recycling of the catalyst or at least the precious

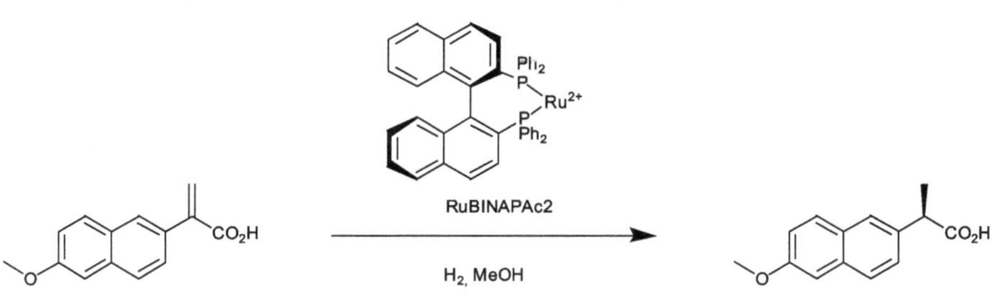

Fig. 3 Synthesis of (*S*)-naproxen using Ru-BINAP (*TTN* = 200, *ee* = 97%)

metal. Other efforts aimed at the heterogenization of catalysts on solid supports or in a non-miscible phase. The latter is especially successful for the industrial-scale production of chemicals in the Rhône-Poulenc/Ruhrchemie hydroformylation process and the oligomerization step in the Shell Higher Olefin Process (SHOP). Generally, immobilization in an additional phase brings forward the problems of mass transfer, especially when the original system already includes gaseous reactants as for hydrogenations and other reactions. Soluble polymers are a viable alternative as scaffolds for homogeneous catalysts [16]. They can be recycled by precipitation or extraction or more elegantly directly by membrane filtration that can be applied for continuous synthesis. This was first demonstrated for enzymes [17, 18] and later extended to man-made catalysts [19, 20].

The concept of covalently binding catalytically active moieties to polymeric soluble supports was initiated by the work of Bayer and Schurig in the mid 1970s [21]. Hydrogenations were among the first reactions carried out and remain popular until today [22]. The Pyrphos ligand introduced by Nagel [23] has been attached to a range of polymeric supports by various groups [24–26].

2 Study Parameters and Results

To establish a model system for reaction engineering of the asymmetric, catalytic hydrogenation, Pyrphos has been selected in this study as an attractive ligand [27], because of its easy access via a short synthesis starting from tartaric acid [28]. The ligand complexes Rh as the precious metal. The enantioselective hydrogenation of 2-*N*-acetyl-acetoamido cinnamic acid was selected as the model reaction used for this catalyst system in this investigation (Fig. 4, *ee* = 94%, quantitative conversion).

The criteria applied in selecting this catalyst are outlined in the following: First, both enantiomers of the ligand are derived conveniently in an off-chiral pool synthesis from tartaric acid. Second, the attachment via the nitrogen in the pyrrolidine ring is possible via amide chemistry and third, the C2-symmetry of the ligand is unaffected by the linkage to the polymer.

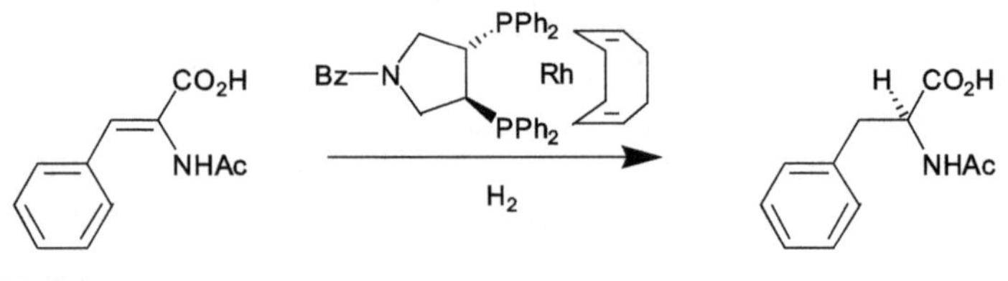

Fig. 4 Enantioselective hydrogenation of 2-acetoamido cinnamic acid catalyzed by Pyrphos (*ee* = 94%, quantitative conversion)

Fourth, modifications to the electronic and steric properties can be carried out. A synthetic route similar to the work of Bergbreiter for the attachment to polymethyl acryl amide was pursued. 4-Amino butyl amide was used as the spacer in order to minimize influence of the polymer backbone.

Aiming for continuous catalysis in multiphase systems combined with membrane filtration introduced difficulties for residence time control of the reactive solvent phase. For continuous membrane filtration a closed liquid volume is favorable because the residence time of the reactants can thus be controlled by flow control of the liquid alone, making mass balancing for the gas phase unnecessary. Differential pressure across the membrane is the driving force for separation by membrane filtration. In a monophasic closed vessel volumetric flow rates of inlet and outlet will be equal if the solvent system is non compressible. Thus, only inlet or alternatively outlet flow must be controlled. As pressure drop across the membrane is hard to predict, a suitable pump will be used to control flow rates independently of pressure [29]. A second phase will pose problems of mass balance if one phase is consumed by either solubility in the mobile phase or reaction. In this case the phase ratio must be controlled by balancing all fluxes in and out of the reactor. In small volumes this is difficult to achieve without variations in volume ratio or hold up volumes, respectively. Presaturation with hydrogen will result in low volumetric productivity as the limiting hydrogen solubility in solvents is generally low [30], as can be determined via the Henry constant.

To overcome this problem we realized a reactor utilizing dense membranes for the continuous supply of hydrogen combined with membrane filtration. Dense membranes can be used for the supply of gases to fluid media. This is known for the oxygen aeration of shear sensitive cell culture broths of mammalian cells and is for example important for the choice of materials for contact lenses as oxygen supply to eye tissue is important for wear stress [31]. The reactors used for cell culture are continuously operated and retention of the cells is usually achieved by immobilization. Residence time or constant reaction volume is controlled by simple overflow. The application of this technology principle was extended to biocatalysis to supply molecular oxygen to reactors containing proteins that are unstable at gas/liquid interfaces [32, 33].

Polymer materials known for their high gas permeability—for example, silicon—are not generally suitable for chemical catalysis as swelling and corrosion will occur with common organic solvents. The excellent chemical resistance of perfluorinated elastomers made them first choice for dense membranes in chemical reactors. Another material for consideration is PEEK (polyether ether ketone) because of its high mechanical strength as well as chemical resistance. Polytetra fluoro ethylene (PTFE) proved to be the best of the commercially available materials in terms of permeability for hydrogen (Fig. 5).

The molar flux can be described in close analogy to heat transfer in a tubular heat exchanger by substituting temperature difference with pressure

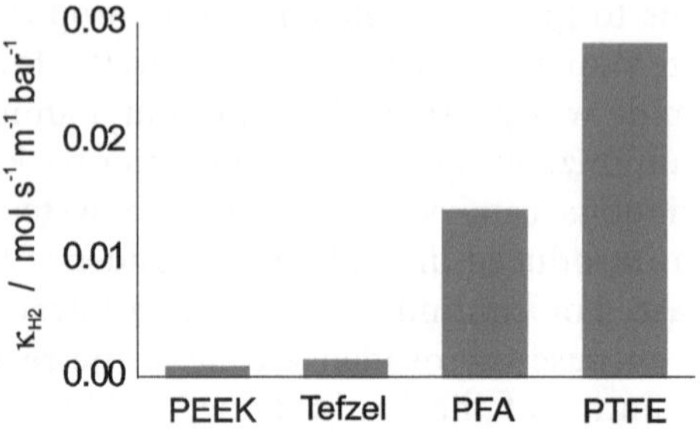

Fig. 5 Hydrogen permeability as a function of material for the mean logarithmic membrane area (tubular membrane: inner diameter 0.79 mm, outer diameter 1.59 m, p_l = ambient, p_i = 11 bar)

difference Δp. The geometry of the tubular membrane is considered by the logarithmic diameter

$$\dot{n} = \frac{\kappa_{H_2} \pi l \Delta p}{\ln d_o - \ln d_i}$$

with the specific transport coefficient k, the length of the tubular membrane l and inner and outer diameter d. Per meter of a tubular PTFE membrane of 1.59 mm (1/16″) outer diameter and 0.79 mm (1/32″) and per bar pressure difference a flux of 1.8 mmol min^{-1} H_2 is obtained. Within the limits of burst pressure a pressure difference of 20 bar can be handled safely. For a flux of 1 mol min^{-1} at 20 bar a length of 27 m is necessary, which when spirally coiled with 10 cm diameter the resulting coil will be 17 cm in length allowing 20% space between the coils. This geometric consideration shows that the hydrogen supply via such a dense membrane is not limiting. For a high surface to reaction volume ratio geometries similar to tubular heat exchangers are advantageous. Concentration gradients in the tubular heat exchanger are circumvented by establishing turbulent flow. The respective minimal flow rates can be calculated for different solvents via the definition of the dimensionless *Reynolds* number *Re*. Turbulent flow is established, if $Re > 2320$.

The utilization of dense membranes shows some advantages compared to standard set ups. Decoupling of the gas and liquid pressures is possible within the limits of the burst pressure of the membrane. Gas volume can be minimized which normally accounts for one third or one half of the total reactor volume. This is an advantage in terms of safety, because lower volumes of potential hazardous hydrogen must be handled. The specific surface area a (area per volume) and the mass transfer coefficient k_l are decoupled (Fig. 6).

The feasibility of the proposed reactor was demonstrated by the hydrogenation of acetamido cinnamic acid methyl ester with a high loading of BBPM/Rh as the catalyst (Fig. 7). Proof of the principle of the new hydrogen supply was performed in a conventional batch reactor with a coiled tubular membrane (180 cm^2) submersed in the liquid phase. Pressure was kept con-

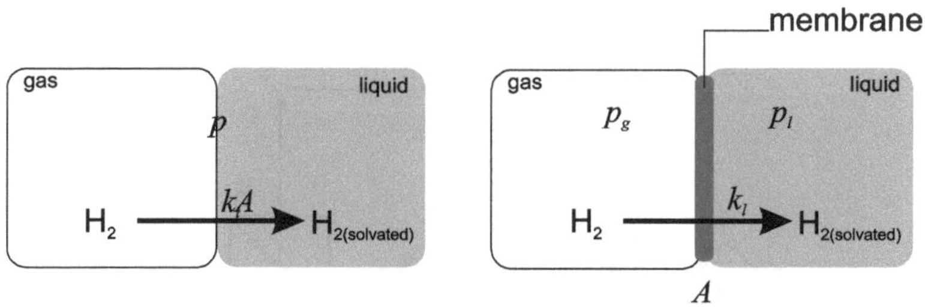

Fig. 6 Decoupling of a and k_l by replacing the interface by a dense membrane (*left*: classical gas/liquid interface, *right*: membrane-separated gas and liquid phase)

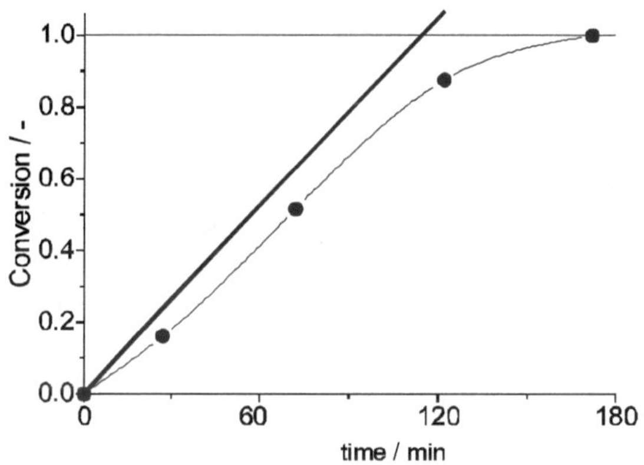

Fig. 7 Conversion as a function of time for the membrane aerated hydrogenation (acetylamidocinnamic acid methyl ester 10 mM, BPPM 1 mM, p_l = ambient pressure, p_g = 9 bar, 2-propanol, 25 °C)

stant and hydrogen uptake was monitored by mass flow metering and the reaction rate closely follows the hydrogen dissemination rate as tracked by independent sampling. The gap in mass balance can be accounted for by the 10 mol % of catalyst activated by hydrogenation itself. A membrane aeration module (400 cm^2) which was incorporated into a loop reactor was set up and successfully tested in batch mode [34].

The proposed membrane aeration was realized as a module that was installed into a loop reactor along with a membrane filtration module (Fig. 8). A tubular PTFE membrane is placed concentrically into a stainless steal tube. The set up is similar to a concentric double-pipe heat exchanger. The design ensures scalability as a tube bundle. The transmembrane pressure drop as the driving force of the permeation of the H_2, can conveniently be controlled by selecting and maintaining the inner pressure of the tubular aeration membrane independently from the pressure within the liquid phase. As a result an effective decoupling of gas and liquid pressure can be achieved within the limits of the burst pressure of the aeration membrane employed. As burst

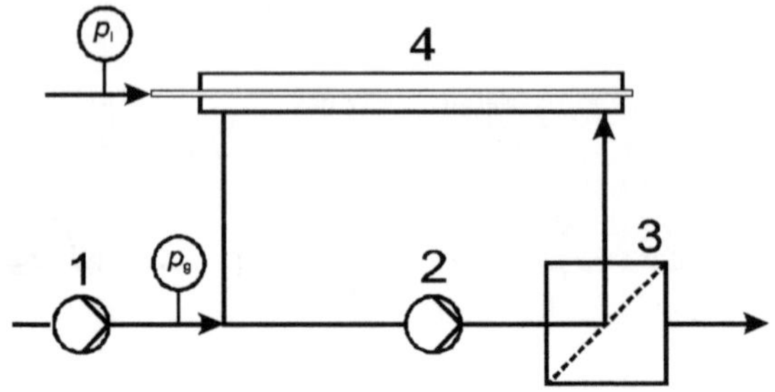

Fig. 8 Flow scheme of the volume aerated membrane reactor (*1*: starting material feeding pump, *2*: circulation pump, *3*: ultrafiltration membrane unit, *4*: volume aeration)

pressure is always higher than kink pressure containing the gas within the innermost tube allows for the highest pressure drop. The flow scheme of the continuously operated reactor for catalytic hydrogenations with macromolecular catalysts is shown in Fig. 8. The filtration module was adopted from previous work [35–40] and serves as a heat exchanger for the reaction loop.

The Pyrphos ligand was chosen as the model system for the continuous hydrogenation. We investigated the kinetics of the reaction in a constant pressure autoclave. Monitoring of hydrogen uptake gives a direct measurement of reaction rates at steady-state conditions or pseudo-zero order in substrate concentration, respectively (Fig. 9). The dependency on catalyst concentration reveals that at low catalyst concentrations the intrinsic properties of the catalyst can be investigated. Interestingly, the reaction rates at low concentrations are unaffected by using THF or methanol as the solvent (Fig. 10). The limiting rate is determined by stirrer speed and the geometry of the autoclave. Overall, a pressure dependent turnover frequency of $2.3 \pm 0.1\ \text{min}^{-1}\ \text{bar}^{-1}$ was determined for the pressure range from 5 to 20 bar. For the polymer-bound Pyrphos preparation a tenfold decrease in activity was found ($0.27 \pm 0.1\ \text{min}^{-1}\ \text{bar}^{-1}$) with unaffected enantioselectivity.

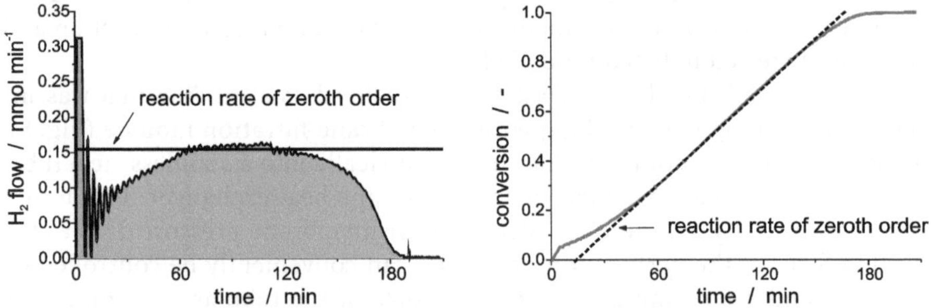

Fig. 9 Typical reaction course of a hydrogenation and resulting pseudo-zero order reaction rate (straight line) (pH_2 = 10.0 bar constant, 0.5 mM Pyrphos, 0.5 M 2-acetoamido cinnamic acid, $T = 25\,°\text{C}$, methanol)

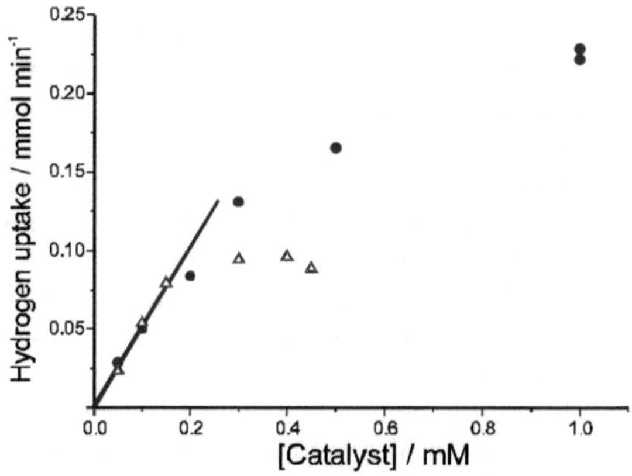

Fig. 10 Hydrogen uptake rate as a function of catalyst concentration for methanol (*circles*) and THF (*triangles*), the *straight line* indicates the calculated turnover frequency ($p = 10$ bar, 0.5 M 2-acetamido cinnamic acid, $\theta = 25\,°C$)

At closer investigation of suitable kinetic models an approach with model discrimination hinted that the Pyrphos catalyst was not stable during the time course of the reaction [41]. The deactivation found was 0.01 per hour leading to a half-life of 58 h for the activity which is to low for continuous reactions. This is a general finding that comparable low loss of catalytically active species which can be tolerated in batch and repetitive batch mode are not acceptable for continuous application. The main reasons for the apparent loss of activity for membrane reactors or multiphase catalysis in general are deactivation or imperfect retention. This can be visualized by plotting residual activity or residual concentration as a function of (dimensionless) time or volume exchanges [42]. Alternatively, the half-life of activity in an ideal

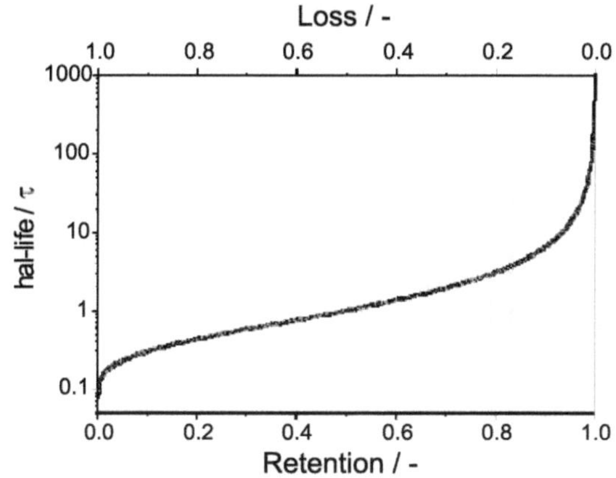

Fig. 11 Half-life as a function of retention or loss, respectively

continuous stirred tank reactor is a function of apparent retention or loss per residence time, due to leaching and/or catalyst deactivation. The half-life is determined as $\ln(2)/\ln(\text{retention}) = \ln(2)/\ln(1 - \text{loss})$ with the residence time $\tau = V_R/F$ and the reactor volume V_R and the volumetric flow rate F. The lower the loss because of permeation the higher is the half-life of activity (Fig. 11). The exponential relationship implies the paramount importance of stability as the limiting factor for continuous applications.

The activation of the homogeneous metal catalyst in continuous reactions can differ greatly from batch experiments, as we found for the addition of base in continuous asymmetric transfer hydrogenation [43]. For hydrogenation catalysts the initiation is generally believed to be the hydrogenation of the protecting diene, for example cyclooctadiene or norbornadiene. In the proposed reactor availability of hydrogen will be limited as the formation of a gas phase in the reaction volume has to be avoided. The choice of protecting diene could therefore have an impact on the transient start of the reaction [44], as catalyst loading in a continuous reactor is high compared to batch reactions where initial effects are small compared to the overall course of the reaction [45]. At higher catalyst concentrations the overall behavior changes and a pseudo-zero order regime cannot be observed. Instead, a prolonged activation phase is observed (Fig. 12). This can be explained by competition of the activation of the catalyst with the hydrogenation reaction itself. As the main reaction is fast due to the high concentration of substrate, the activation is starved of hydrogen if mass transfer is limited. However, treatment of the Pyrphos catalyst precursor bound to cyclooctadien with hydrogen under reaction conditions did not yield a higher initial activity. Initial activation or lag phase, respectively is observed nonetheless. To check for full activation of the catalyst the reaction was continued by subsequent further addition of substrate (Fig. 13). In the second part no further activation was observed and the rate was equal within the experimental error. Activation may therefore be a combination of ligand exchange and liberation of the ac-

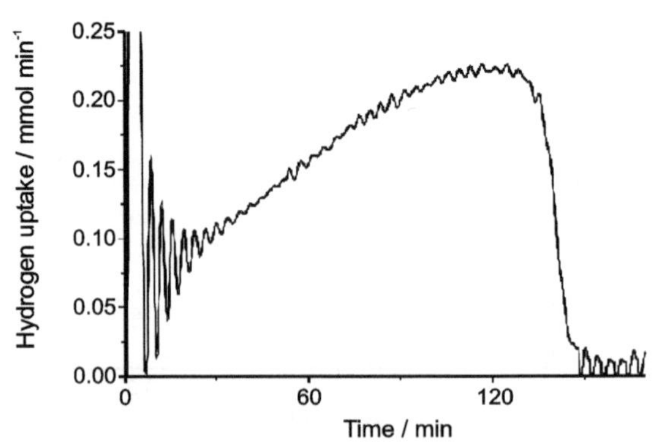

Fig. 12 Hydrogen uptake rate as a function of time for high catalyst loading

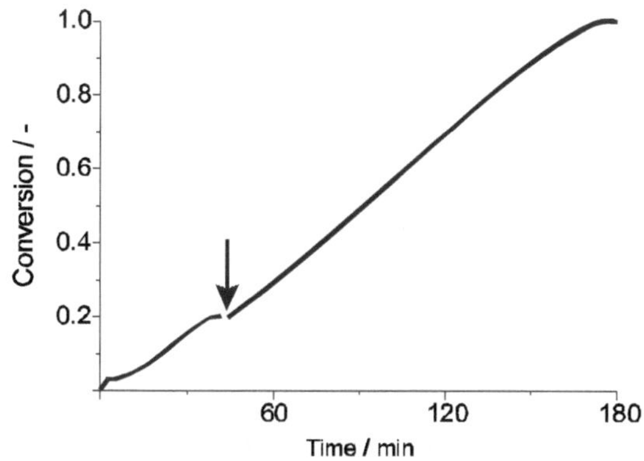

Fig. 13 Conversion vs. time with addition of substrate solution (*arrow*)

tive catalyst by hydrogenation of the protecting diene. These findings are in accordance with activation behavior of DIOP [46].

The continuous process was carried out in a volume aerated membrane reactor with minimized hold-up volume. The reactor was successfully used on the example of the continuous enzyme catalyzed reduction of $NADP^+$ to NADPH with a hydrogenase from *Pyrococcus furiousus* [47]. A total turnover number (TTN) of more than 66 000 and a space time yield of 130 g L^{-1} d^{-1} was obtained. The experimental setup was slightly extended and selectivity towards the intermediate product NADPH was optimized [48].

The synthesis of the Pyrphos chemzyme applied for the continuous synthesis was carried out via a convergent synthesis in close analogy to published protocols [49, 50]. The Pyrphos ligand was covalently bound to a poly(*N*-*iso*-propyl-methacrylamide). Furthermore, other polymer-enlarged Pyrphos systems have also been described in the literature [51, 52]. The advantage of this particular chemzyme is the additional spacer unit, 4-amino butyl amide, in order to prevent interactions or interference with the polymeric backbone thanks to the increased distance. In contrast to other polymer-coupled Pyrphos systems, the selectivity of this polymer-bound system is not changed with the preparation of this chemzyme relative to its low-molecular relative. Figure 14 shows the conversion as a function of the number of residence times for a continuous experiment. At a residence time of $\tau = 2.0$ h the hydrogen supply via the PTFE membrane seems to be limiting, and only a low conversion is reached. By increasing the residence time to $\tau = 5.7$ h, the conversion is approximately doubled. A further addition of catalyst after 97 h (17 residence times of 5.7 h) leads to a maximal conversion of 0.6, which is instable. Furthermore, a fast decrease of the conversion is observed. Non-linear regression of the conversion decrease delivers an apparent retention of $r = 0.78$ for the chemzyme. The dotted line denotes the case, where no retention of the chemzyme would be given. Apparently the catalyst is not stable under reaction conditions. The independently determined retention of 0.99 and the

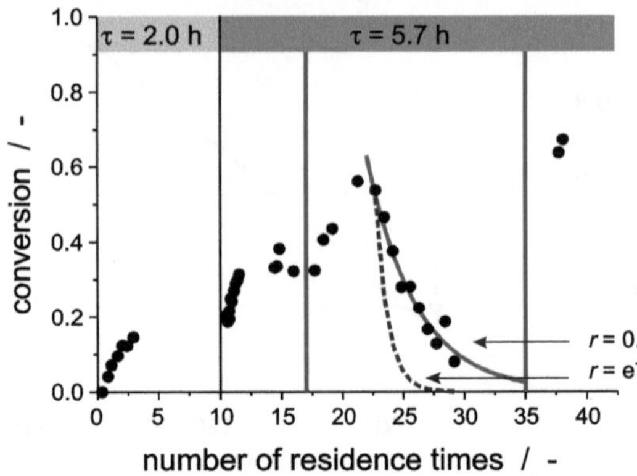

Fig. 14 Hydrogenation in the continuously operated volume aerated membrane reactor with the Pyrphos chemzyme ($p_l = 9.0$ bar, $p_{H_2} = 24.0$ bar constant, 0.1 M 2-acetoamido cinnamic acid, $T = 40\,°C$)

deactivation predicted from modeling reaction kinetics do not sufficiently cover the magnitude of apparent loss of activity.

3
Conclusion

Asymmetric hydrogenation with homogeneously soluble catalysts, either metal or bio-catalysts, is without doubt a key technology for asymmetric synthesis. Membrane aeration combined with membrane filtration is a means for the convenient continuous application of macromolecular catalysts for hydrogenation with molecular hydrogen. To enable continuous application a high catalyst robustness and apparent stability have to be given to justify the necessary technical and in the case of the chemzyme synthetical efforts.

Membrane aeration allows decoupling of process parameters of pressures and mass transfer. Furthermore, efficient design allows for minimization of the non-reactive gaseous volume.

Acknowledgements We thank Andreas Franz, Daniela H. Müller and Christian R. Reimers (all of Forschungszentrum Jülich) for skilful technical support. The research was made possible by financial support from the German Federal Ministry of Education and Research (BMBF) and Degussa AG.

References

1. Osborn JA, Jardin FH, Young JF, Wilkinson G (1966) J Chem Soc, p 1711
2. Knowles WS, Sabacky MJ (1968) Chem Comm, p 1445
3. Kagan HB, Dang T-P (1972) J Am Chem Soc 94:6429

4. Tang WJ, Zhang XM (2003) Chem Rev 103:3029 and references cited therein
5. Brown JM, Chaloner PA (1980a) J Chem Soc-Chem Commun, p 344
6. Brown JM, Chaloner PA (1980b) J Am Chem Soc 102:3040
7. Halpern J (1982) Science 217:401
8. Achiwa K (1976) J Am Chem Soc 98:8265
9. Nagel I (1984) Angew Chem 96:425
10. Burk MJ, Feaster JE, Nugent WA, Harlow RL (1993) J Am Chem Soc 115:10125
11. Pye PJ, Rossen K, Reamer RA, Tsou NN, Volante RP, Reider PJ (1997) J Am Chem Soc 119:6207
12. Togni A, Dorta R, Kollner C, Pioda G (1998) Pure Appl Chem 70:1477
13. Muniz K, Bolm C (2000) Chem Eur J 6:2309
14. Noyori R (1994) Asymmetric Catalysis in Organic Synthesis. Wiley, Tokyo
15. Koten Gv, Leeuwen PWv (1999) In: Santen RVv, Leeuwen PWv, Mouljin JA, Averil BA (eds) Catalysis: An integrated approach, Vol. 123. Elsevier
16. Bayer E, Schurig V (1975) Angew Chem 87:484
17. Lütz S, Rao N, Wandrey C (2005) Chem Ing Tech 77:1669
18. Kula M-R, Wandrey C (1987) Methods in Enzymology 136:9
19. Kragl U, Dwars T (2001) Trends in Biotechnology 19:45
20. Wöltinger J, Drauz K, Bommarius AS (2001) Appl Cat A 221:171
21. Mecking S (2005) In: Cornils B, Herrmann WA, Horváth IT, Leitner W, Mecking S, Olivier-Bourbigou H, Vogt D (eds) Multiphase Homogeneous Catalysis. Wiley, Weinheim, p 753
22. Greiner L, Liese A (2006) In: Cornils B, Herrmann WA, Horvath IT, Leitner W, Mecking S, Olivier-Bourbigou H, Vogt D (eds) Multiphase Homogeneous Catalysis. Wiley, Weinheim, p 819
23. Nagel U (1984) Angew Chem 96:425
24. Malmström T, Andersson C (2000) J Mol Cat A 157:79
25. Engel GD, Gade LH (2002) Chem Eur J 8:4319
26. Bergbreiter D (2000) In: De Vos DE, Vankelecom IFJ, Jacobs PA (eds) Chiral Catalyst Immobilization and Recycling. Wiley, Weinheim
27. Nagel U, Kinzel E (1986) Chem Ber 119:1731
28. Beck W, Nagel U (1988) EP 0151282
29. Laue S, Greiner L, Wöltinger J, Liese A (2001) Adv Synth Catal 343:711
30. Fogg PC, Gerrard W (1991) Solubility of gases in liquids: A critical evaluation of gas/liquid systems in theory and practice. Wiley, Chichester, UK
31. Compan V, López ML, Andrio A, López-Alemany A, Refojo MF (1999) J Appl Polym Sci 72:321
32. Rissom S, Schwarz-Linek U, Vogel M, Tishkov V, Kragl U (1997) Tetrahedron: Asymmetry 8:2523
33. Bongs J, Hahn D, Schörken U, Sprenger GA, Kragl U, Wandrey C (1997) Biotechnol Lett 19:213
34. Greiner L, Müller DH, van den Ban ECD, Wöltinger J, Wandrey C, Liese A (2003) Adv Synth Catal 342:679
35. Laue S, Greiner L, Wöltinger J, Liese A (2001) Adv Synth Catal 343:711
36. Giffels G, Beliczey J, Felder M, Kragl U (1998) Tetrahedron: Asymmetry 9:691
37. Wöltinger J, Bommarius AS, Drauz K, Wandrey C (2001) Org Proc Res Dev 5:241
38. Kragl U, Dreisbach C (1996) Angew Chem Int Ed Engl 35:642
39. Kula MR, Wandrey C (1987) Meth Enz 136:9
40. Iding H, Siegert P, Dünnwald T, Müller M, Greiner L, Liese A, Grötzinger J, Demir A, Pohl M (2000) Chem Eur J 6:1483

41. Greiner L, Brik Ternbach M (2004) Adv Synth Catal 346:1392
42. Mecking S (2005) In: Cornils B, Herrmann WA, Horváth IT, Leitner W, Mecking S, Olivier-Bourbigou H, Vogt D (eds) Multiphase Homogeneous Catalysis. Wiley, Weinheim, p 753
43. Greiner L, Laue S, Liese A, Wandrey C (2006) Chem Eur J 12:1818
44. Börner A, Heller D (2001) Tetrahedron Lett 42:223
45. Cobley CJ, Lennon IC, McCague R, Ramsden JA, Zanotti-Gerosa A (2001) Tetrahedron Lett 42:7481
46. Nindakova LO, Shainyan BA (2001) Russ Chem Bull 50:1855
47. Greiner L, Müller DH, van den Ban ECD, Wöltinger J, Wandrey C, LieseA (2003) Adv Synth Catal 342:679
48. Mertens R, Wandrey C, Liese A (2005) Chem Ing Tech 77:609
49. Bergbreiter DE, Case BL, Liu U-S, Caraway JW (1998) Macromolecules 31:6053
50. Bergbreiter DE (2002) Chem Rev 102:3345
51. Malmström T, Andersson C (1999) J Mol Cat A: Chemical 139:259
52. Malmström T, Andersson C (2000) J Mol Cat A: Chemical 157:79

Top Curr Chem (2007) 276: 125–154
DOI 10.1007/128_2007_119

Published online: 31 March 2007

Exploiting Nuclear Spin Polarization to Investigate Free Radical Reactions via in situ NMR

Lars T. Kuhn · Joachim Bargon (✉)

Institute of Physical & Theoretical Chemistry, University of Bonn, Wegelerstrasse 12, 53115 Bonn, Germany
bargon@uni-bonn.de

Abstract In situ NMR spectroscopy can be applied to investigate chemical reactions during which free radicals occur as intermediates. In chemical systems of low molecular weight, nuclear spin polarization results from the spin selectivity of free radical reactions because a pair of radicals, like any other given set of particles, has to obey the exclusion principle. Therefore, this system reacts selectively in terms of the participating nuclear spins when forming a chemical single bond. As a consequence, strong transient absorption and emission lines occur in NMR spectra acquired during a reaction of free radicals. This extraordinary phenomenon has become known as chemically induced dynamic nuclear polarization (CIDNP). Ever since its experimental discovery and theoretical verification, CIDNP has been employed to study the mechanisms of free radical reactions in solution. As such it has proven to be a very valuable tool for the elucidation of the mechanism of these reactions and, more importantly, to discriminate reaction pathways that include the formation of transient radical species from those that exclusively follow a "diamagnetic" route, i.e., a pathway where no paramagnetic intermediates are formed whatsoever. More recently, the photo-CIDNP technique has also been employed extensively to probe the surface-accessibility of aromatic amino acid side-chains bound within a protein. As such, it can be used to study the dynamic features of a protein during folding, refolding, and also in the equilibrium or "steady state", yielding both qualitative and quantitative information.

This review outlines the historical development of the CIDNP technique as well as its theoretical background. This is followed by a series of examples showing how CIDNP can be used to elucidate reaction pathways of chemical transformations comprising diamagnetic intermediates. Additionally, we present examples of how "biological" CIDNP experiments are usually performed and we show what kind of information can be extracted from these studies.

Keywords Chemically induced dynamic nuclear polarization · CIDNP · NMR · Photochemistry · Radical reactions

Abbreviations

CIDEP	Chemically induced dynamic electron polarization
CIDNP	Chemically induced dynamic nuclear polarization
CINOE	Chemically induced nuclear Overhauser effect
CW	Continuous wave
DBPO	Dibenzoylperoxide
FT	Fourier transformation
NMR	Nuclear magnetic resonance
NOE	Nuclear Overhauser effect
FID	Free induction decay
RPM	Radical pair mechanism
ESR	Electron spin resonance
EPR	Electron paramagnetic resonance
TEA	Triethylamine

1 Introduction

Ever since its accidental discovery [1–3] in the form of intense emission and enhanced absorption lines in NMR spectra about 40 years ago, the nuclear spin polarization phenomenon termed "chemically induced dynamic nuclear polarization" or simply CIDNP has been used to investigate a great variety of free radical reactions. Especially attractive is its potential to determine structural details of biochemically important molecules, which has been demonstrated convincingly for proteins [4–6].

The observation of CIDNP requires in situ NMR spectroscopy, which allows the investigation of chemical reactions during which free radicals occur as intermediates. At least in chemical systems of a low molecular weight, nuclear spin polarization results from the spin selectivity of free radical reactions, because a pair of radicals has to obey the exclusion principle when forming a chemical single bond. In principle, the CIDNP phenomenon is also suited for the identification of both the nature of intermediate radicals as well as their reaction products in macromolecular systems. Observing CIDNP in chemical systems of high molecular weight, however, is intrinsically more difficult, chiefly because of the very short relaxation times of the nuclei within or even in the mere presence of macromolecules. This is especially true for situations where the reaction products themselves have a high molecular weight, and it is even worse in cases where the product molecules contain long molecular chains, e.g., alkyl chains. The situation is, however, not totally hopeless, provided that the NMR equipment is adapted to suit the required boundary conditions for investigating such reactions.

CIDNP is an indirect method for observing and characterizing free radicals via the polarization observed in the corresponding reaction products. Since the extent of CIDNP is stronger the more short-lived the participating radicals, this spectroscopic method complements direct methods of investigating free radicals, such as ESR (electron spin resonance) or EPR (electron paramagnetic resonance) both in its applicability as well as in the type of the results that are obtained. The magnitude of CIDNP enhancements in reaction products of low molecular weight is governed by the ratio of the magnetic moments of the electron versus those of the nuclei being observed. Therefore, for protons the experimentally observed enhancements typically range in regions of one to almost three orders of magnitude; for heteronuclei this enhancement will accordingly be even higher.

The related CIDEP effect in ESR spectra is much smaller than typical CIDNP enhancements in NMR spectra. The CIDEP enhancements are of the order of unity in steady state spectra, but can reach values of up to two orders of magnitude. To detect these higher CIDEP enhancements, ESR spectrometers must be modified to allow observation of the radicals within microseconds of their formation.

2 Chemically Induced Dynamic Nuclear Polarization (CIDNP)

2.1 Experimental Aspects of the CIDNP Phenomenon

In order to observe and acquire CIDNP spectra no modification of conventional NMR spectrometers operating in either Fourier transform (FT) or continuous wave (CW) mode is required. The CW mode used to be typical for NMR spectrometers prior to the advent of the FT technology, which requires computers to generate the NMR spectrum starting from a free induction decay (FID).

Photochemical generation of free radicals inside the NMR probe requires appropriate attachments to conventional NMR spectrometers. These days, appropriately modified probes are normally available commercially, whereby the required light is admitted either along the axis or from the side of the spinning NMR tube. In the latter case, mirrors may be used to deflect and admit the radiation through an appropriately shaped and spaced radio frequency coil. However, the required modifications can be minor, especially if an appropriately shaped UV light rod made from fused quartz is used. Such a rod, typically with an outer diameter between 4 and 10 mm (depending on the inner diameter of the NMR tube), extends into the open spinning sample from the top with its tapered end which extends ca. 1 cm into the liquid sample. There, it introduces efficient stirring agitated by the spinning in addition to allowing UV irradiation of the continuously stirred solution. This stirring effect is especially beneficial in ^{13}C-FT-NMR spectrometers and has merit even without accompanying UV illumination. Recording of CIDNP spectra with FT-NMR spectrometers in combination with a pulsed light source (for example a laser) has certain advantages over a CW spectrometer in combination with a Hg – Xe arc lamp. This is because the effects of NMR relaxation can be greatly reduced or eliminated. NMR relaxation puts a major limitation on the accuracy of the ESR parameters that may be determined from CIDNP spectra.

More recent approaches of guiding light from a laser source into the coil region of the NMR spectrometer make modifications of the NMR probe completely unnecessary. In these cases laser light is guided directly via an optical fiber from the laser into the NMR probe of the spectrometer. The loss of light intensity using this setup is negligible and a compromise in field homogeneity due to the presence of the fiber tip close to the coil region can easily be achieved using modern "gradient shimming" procedures [7].

2.2 Spin Chemistry

The CIDNP phenomenon can readily be interpreted in terms of the radical pair mechanism, where very small magnetic energies can change nonequilibrium spin interconversion processes. The progress of both CIDNP and the analogous phenomenon termed "chemically induced dynamic electron polarization" (CIDEP) [8] has brought about the advent of a new branch of chemistry, which is called "spin chemistry". Spin chemistry means that the chemical options of a pair of free radicals depend upon the relative alignment of the spins of the two unpaired electrons and hence on the combined multiplicity of the radical pair.

Broadly defined, spin chemistry deals with the effects of electron and nuclear spins in particular, and magnetic interactions in general, on the rates and yields of chemical reactions. The consequences of these investigations are not only manifested as spin polarization in ESR and NMR spectra, but also in the magnetic field dependence of certain chemical processes. Applications include studies of the mechanisms and kinetics of free radical reactions in solution, the energetics of photosynthetic electron transfer reactions, enzyme catalysis, and various magnetokinetic effects. These include possible biological effects of extremely low frequency and radio frequency electromagnetic fields on the yields of chemical reactions [9], the mechanisms by which it is assumed that animals can sense and use the Earth's magnetic field for orientation and navigation [10–12], and the possibility of manipulating radical lifetimes so as to control the outcome of their reactions.

2.3 The Radical Pair Mechanism – A Qualitative Description

The radical pair mechanism (RPM) [13–19] accounts for the generation of CIDNP via a nuclear spin state-dependent reaction between the two partners of a radical pair. Free radicals contain one unpaired electron, and hence they represent *doublet* species with respect to their electron spin state. Pair formation of two radicals leads to two possibilities for the total electron spin of the pair, namely either the singlet state ($S = 0$) so the corresponding multiplicity is $2S + 1 = 1$ (i.e., a *singlet*) or the triplet state, whereby $S = 1$ and the multiplicity is $2S + 1 = 3$ (i.e., a *triplet*).

The ability of the two radicals to react with each other depends on the combined spin multiplicity (singlet or triplet) of the two unpaired radical electrons. A transition between the non-bonding (triplet) and the *reactive* (singlet) electron spin state (a so-called singlet–triplet transition) depends on interactions between the unpaired electrons of the radicals and any magnetic influences, such as the small nuclear magnetic moments of the surrounding nuclei in a molecule or an external magnetic field B_0. The way in which these interactions affect a singlet–triplet transition depends on the nuclear spin states of the magnetic nuclei within the radicals. The reactivity of the electron spin state also depends on the type of bond being formed.

Certain nuclear spin states are more likely to cause a transition to the reactive electron spin state; therefore, reaction products derived from these very nuclear spin states have a greater probability of formation. For a large number of radical pairs, this leads to a non-Boltzmann distribution of nuclear spin state populations in their reaction products, thereby giving rise to the CIDNP phenomenon observed in NMR spectra recorded in situ. However, the nuclear spin-dependent singlet–triplet transition that ultimately gives rise to CIDNP only occurs once the two partner radicals have moved far enough apart so that the exchange interaction between the two radicals is no longer dominant but on the order of the hyperfine coupling interaction.

2.4 Chemical Dynamics of CIDNP

The dynamic process that leads to CIDNP may also be explained with the aid of a simplified graph of the potential energy of the reactive and non-reactive electron spin states of a radical pair as a function of radical separation. From the potential curves the reaction path of a radical pair can be derived. The details of this path, however, are dependent on whether one uses a *diffusion* model or an *exponential* model.

In such a picture the lower curve usually represents the reactive electron spin state, which is normally assumed to be the singlet state S. The upper curve represents the non-reactive triplet state. In a strong magnetic field, however, the triplet state is split into three components. These three components interconvert into each other owing to fast relaxation rates. J is the exchange integral. For short distances ($r < r'$), the exchange interaction splits the singlet and triplet levels, but for larger distances ($r > r'$), the two states become degenerate as the exchange interaction disappears. True singlet and triplet states are defined only for a coupled set of two electron spins, such as occur in the region where $r < r'$. In this region, the spins couple via the exchange interaction. For $r = d$ (i.e., the bond distance) the exchange integral J assumes its largest value, corresponding to the bonding singlet state $^1(R_1R_2)$ and the antibonding triplet state $^3(R_1R_2)$. As r increases to r', radical pairs form whose exchange interaction is much smaller, but they are still either in the singlet state or the triplet state. For distances $r > r'$, the two radicals are so far apart that they no longer interact ($J = 0$) in which case the singlet and triplet states are not defined. Rather, each radical behaves independently of the other; therefore, each resides in a doublet state $^2(R_1^{\cdot}) +^2 (R_2^{\cdot})$.

In the presence of a strong magnetic field several aspects of the aforementioned situation change. One aspect that changes is that singlet and triplet states are defined with respect to a new axis of quantization, namely the external magnetic field B_0. This means that singlet and triplet states may also be defined for $r > r'$, where the exchange integral is zero, even though the spins of the unpaired radical electrons are not coupled to each other. Another aspect that changes is that the Zeeman interaction splits the three components of the triplet state into three different energy levels; this is because each state of the triplet possesses a different magnetic quantum number (1, 0, or − 1). When the exchange interaction between the two partner radicals is small or zero, only the T_0 state remains degenerate with the singlet state; therefore, virtually only the T_0 state participates in singlet–triplet transitions in a strong external magnetic field B_0.

Since CIDNP is usually investigated within an NMR spectrometer, a strong magnetic field is present, and this is the case assumed for the theory and experiments described here. A singlet–triplet transition occurs at a distance where the exchange interaction is small enough to allow other forces to dominate. These other forces are the previously mentioned interactions of the unpaired electrons with magnetic nuclei and the external magnetic field. These perturbations can now cause a singlet–triplet transition, because the energy separation of the two states is no longer insurmountable.

Based on these considerations, the reaction path of a radical pair in a strong magnetic field is now examined schematically. First, by some experimental means, a radical pair is induced to form either in the singlet or triplet state. Because of random molecular motion, the distance between the two radicals is time-dependent; moreover, the separations achieved are large enough to allow a singlet–triplet transition. One factor affecting this transition is due to the nuclear spins of the magnetic nuclei, which couple with the electron spins of the respective radicals. If the radical pair starts in the non-reactive triplet state, certain nuclear spins favor a transition to the reactive singlet state. This singlet state radical pair can now react to either form the nuclear spin polarized recombination product $^1(R_1R_2)^*$, other singlet state combination products, singlet state disproportionation products, or some assortment thereof.

The combination and disproportionation products have a greater probability of having nuclear spin states that favor a singlet–triplet transition. A radical pair, whose nuclear spins render it unfavorable for a singlet–triplet transition, remains in the triplet state and is more likely to separate and yield two free radicals. These two free radicals may now attack the solvent by abstracting hydrogen, chlorine, bromine, or the like to form an *escape product*. Alternatively, the two free radicals themselves may come under attack by various scavengers and form *scavenger products*. These escape and scavenger products (as opposed to the aforementioned combination and disproportionation products), therefore, have a greater probability of possessing those nuclear spin states that do not favor a singlet–triplet transition. These escape and scavenger product reactions, however, are themselves not nuclear spin-selective; rather they happen immediately, allowing no time for a nuclear spin-dependent singlet–triplet transition, but merely reflect the nuclear spin polarization results of an earlier singlet–triplet transition. Analogous pathways may be traced for cases where the precursor is the singlet state or the exit channel is the triplet state. For more than one radical pair, both the precursors and exit channels may be a mixture of singlet and triplet states.

With respect to radical separation and recombination, the principles of the diffusion model [20, 21] best account for the resulting CIDNP spectrum; hence, this model provides good quantitative results. In the diffusion model, the two partner radicals of the radical pair diffuse apart after their initial contact, and then the radicals undergo a series of reencounters (Fig. 1). This gives the radical pair sufficient time to undergo a singlet–triplet transition and hence sufficient opportunity to react. This type of diffusion modeling has been used extensively by Noyes [22] in a treatment of classical reaction kinetics, and it was introduced into CIDNP theory by Adrian. The exponential model, as applied in the initial formulation of the RPM [13], on the other

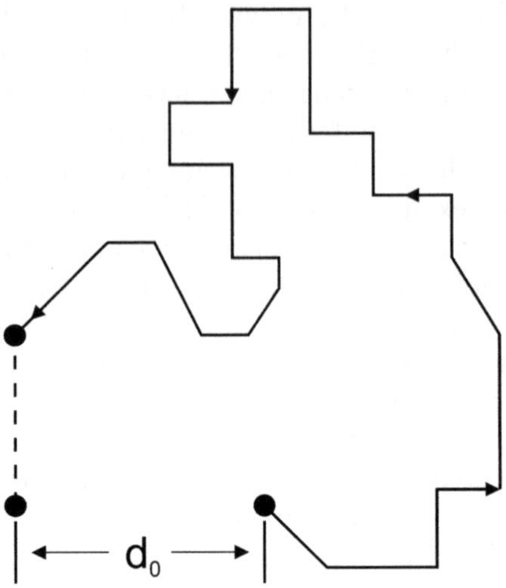

Fig. 1 Formation and separation of radical pairs according to the diffusion model. The distance d_0 denotes the initial separation of the two radicals prior to their diffusion

hand, assumes that the radical pair is retained within a solvent cage. While this model successfully predicts many of the qualitative aspects of absolute CIDNP intensities, better quantitative results for absolute CIDNP intensities come from the diffusion model.

2.5 Electron Spin State Reactivity

The RPM theory of CIDNP usually assumes that the bond to be formed from the interaction of the two free radicals in a radical pair is a single σ bond. However, this theory has been expanded to include the formation of π bonds as well. For example, an electron transfer reaction that moves an electron from a radical anion to a radical cation may result in the formation of a π bond [23, 24]. Whereas σ bond formation arises only from the singlet state of a radical pair, π bond formation can arise from its singlet *or* its triplet state.

The reason behind singlet state reactivity in the formation of a σ bond is based upon an assumed similarity between the chemical bond in the hydrogen molecule and chemical bonds in other molecules, which are more complicated. This can be illustrated by plotting the potential energy of the bonding and anti-bonding states of the hydrogen molecule as a function of internuclear distance. The bonding and anti-bonding states, which are governed by quantum mechanical exchange forces, are correlated with the total electron spin of a radical pair by the exclusion principle. Thus, the repulsive branch of the potential energy curve corresponds to the triplet state of a radical pair, and the attractive branch of the potential energy curve corresponds to the singlet state of a radical pair. These curves are assumed to be true for more complicated radical pairs. From this graph it is apparent that the triplet state experiences a repulsive force for short distances, while the singlet state experiences an attractive force down to an equilibrium distance ($r = d$), beyond which strong repulsive forces set in ($r < d$). Hence, a σ bond between two radicals can be formed only if the electrons of the radicals are in the singlet state; otherwise, the repulsive force of the triplet state prevents the radicals from combining.

For electron spin state reactivity in π bond formation, the treatment is similar except that the triplet state may also have a potential energy minimum. If this is the case, the triplet state can also experience an attractive force down to an equilibrium distance, thereby resulting in π bond formation. Thus, π bonds may form from the singlet and triplet state, depending on the exact nature of the system.

2.6
Spin Selectivity and Magnetic Field Dependence of Chemical Reactions

As pointed out before, it is a fundamental principle of radical reactions that the multiplicity of the two unpaired electrons that are supposed to form a chemical single bond remains unchanged during "elementary chemical events". The same holds true for their projections. This law of the conservation of the total spin results in an important consequence: *chemical reactions are spin selective*, i.e., they are only allowed for such spin states of products whose total spin is identical to that of the reagents, but strictly forbidden if they require a spontaneous spin change.

Accordingly, the chemical coupling in a pair of radicals, which comprises two options (either combination or disproportionation reactions), is "spin allowed" only for singlet pairs. Triplet radical pairs may not react until their spin state changes to the singlet state. Thereupon the reaction starts generating the product molecules in their singlet state. Likewise, in the simplest case, the formation of H_2 from two hydrogen atoms is only permitted for singlet encounters, and the same is true for the formation of ethane from two methyl radicals.

Some biologically important chemical reactions are also spin-selective and some of those are even sensitive to the presence of a small magnetic field. Most of all, this concerns certain processes in photosynthesis. Indeed, many important details of these reactions have by now been elucidated exploiting their magnetic field dependence. Likewise, at the level of cell biology and phenomenological enzyme chemistry there have been many observations of an influence of a magnetic field on the efficiency of biological processes catalyzed by enzymes, such as the determination of the ratio between oxidation and oxidative phosphorylation, or the activity of peroxidase and other enzymes, to name just a few.

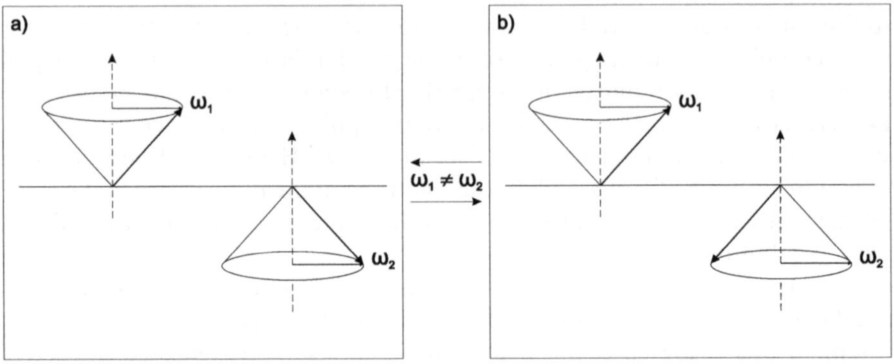

Fig. 2 Multiplicity and spin-selective reactivity of radical pairs. **a** Triplet pair: the two radical spins precess "out of phase" and product formation is forbidden; **b** Singlet pair: "in phase" precession allows the formation of recombination products. ω_1 and ω_2 denote the individual precession frequencies of the two radicals in the radical pair, respectively

Since triplet radical pairs, which are actually formed predominately upon a statistical encounter of two free radicals, are not persistent, they cannot be completely *inert*. Therefore, a process must exist that converts them from their initial triplet multiplicity to that of the chemically highly reactive singlet counterpart. The radical pair theory of the CIDNP phenomenon [13] explains this process to be the result of a dephasing process (Fig. 2).

2.6.1 Basic Features of the CIDNP Method

According to the aforementioned radical pair theory of CIDNP, the spin selectivity of radical reactions accounts for the observation of strongly enhanced absorption and emission lines during in situ NMR observations. In the presence of an external magnetic field B_0, the two unpaired electrons of the partner radicals of the pair precess with a frequency that is proportional to the strength of the magnetic field. Among the possible combinations of their precessing spins, there exists one, in which the spins are oriented opposite to each other, whereby their individual precession occurs exactly 180° "out of phase".

This arrangement represents the singlet state since this is the only combination with a resulting magnetic moment of zero. All other remaining arrangements, a total of three, have a triplet multiplicity. Their individual electron spins are all aligned parallel to each other, and depending on the orientation of their projections on the axis of the magnetic field, they are referred to as the "up" (T_{+1}), "down" (T_{-1}), or "perpendicular" (T_0) sublevel of the triplet manifold. These three species reflect the three allowed orientations of a system with a resulting electronic spin of one. Initially, the phase relation of the individually precessing electron spins within these triplet pairs is "in phase" in all three cases. Each state of the triplet manifold has a non-zero magnetic moment. Since they are of the same multiplicity, they exchange efficiently and hence rapidly among each other. Essentially, therefore, the only difference between the singlet and the triplet state with the antiparallel electron spins can be reduced to a difference in the phase relationship of the precessing electron spins in the pair.

A system consisting of two unpaired electrons, 1 and 2, in two different molecular orbitals with two possible orientations (α and β) relative to an external magnetic field B_0, can be found in any of the possible spin states defined by S and m_S. Accordingly, this adds up to a number of four different spin states, namely $\alpha_1\alpha_2$, $\alpha_1\beta_2$, $\beta_1\alpha_2$, and $\beta_1\beta_2$ (Fig. 3).

Since the spin states interact with each other the two initial $S = 1/2$ states have to be combined to $S = 0$ and $S = 1$ states, respectively, i.e., to a singlet and to three triplet states with the following coefficients and representations:

$$S_0: \ |0;0\rangle = \frac{1}{\sqrt{2}}\left(\alpha_1\beta_2 - \beta_1\alpha_2\right) ; \quad T_{\pm 0}: \ |1;0\rangle = \frac{1}{\sqrt{2}}\left(\alpha_1\beta_2 + \beta_1\alpha_2\right) ;$$
$$T_{-1}: \ |1;-1\rangle = \beta_1\beta_2 ; \quad T_{+1}: \ |1;+1\rangle = \alpha_1\alpha_2 .$$

If the two radical electrons are magnetically equivalent, i.e., if the two radicals have the same g value, their phase relation will remain constant, as long as any possible interaction with magnetically active nuclei can be ignored. Their initial phase relationship will change, however, if either the radicals are chemically different, i.e., if they have different g values, or if the magnetically active nuclei in otherwise chemically identical radicals induce a change of the electronic multiplicity, i.e., a singlet–triplet multiplicity change. The associated dephasing process is called *singlet–triplet mixing*. It controls both the types of chemical products that may result, and the details of the nuclear spin polarization with which the products are initially born (Fig. 4).

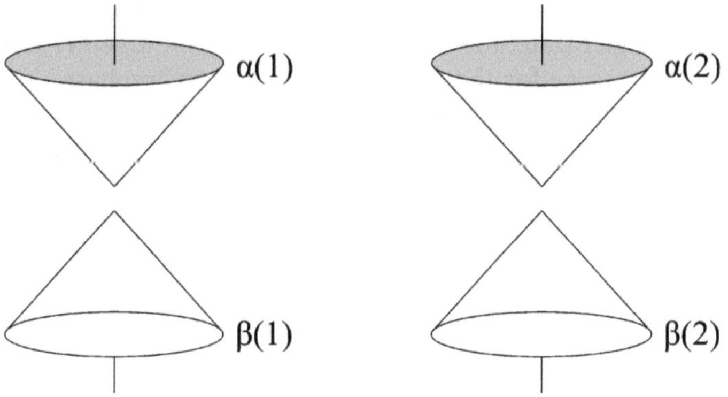

Fig. 3 Vector model representation of the four possible (allowed) states for two non-interacting spins

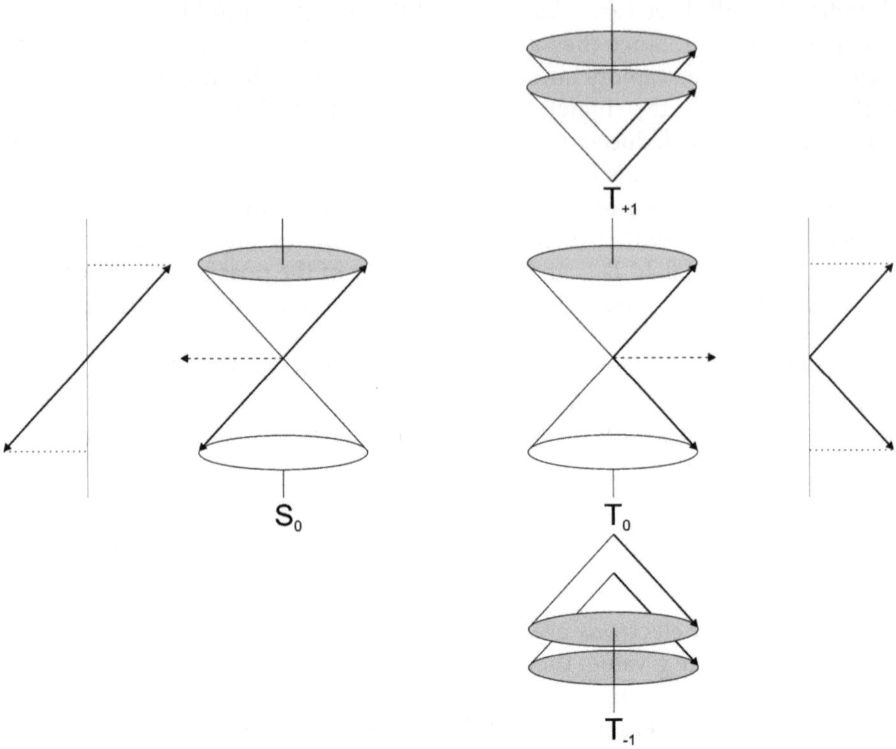

Fig. 4 Vector model of the singlet and triplet states in radical pairs. The net magnetic moment of the singlet state is zero and the triplet manifold is characterized by non-zero magnetic moments

Once the two radicals are formed in solution as the consequence of a bond cleavage, they remain close together for a short time, held in place by a "cage" of solvent molecules. In order for the radicals to combine, the two electrons must be aligned favorably, i.e., they must mate in the singlet state. If they are not in the singlet state, the radicals have to remain together long enough within the solvent cage for the triplet multiplicity to "rephase" and thereby intersystem-cross into the singlet state. If this singlet–triplet mixing takes too long, the radicals may escape the solvent cage before rephasing occurs and thereafter engage into some other types of reactions, forming so-called escape products. Accordingly, recombination products are formed predominantly from the fraction of triplet radical pairs that contain nuclear spins favoring the required conversion to the singlet electronic state. Escape products will be formed from such radical pairs with nuclear spins unfavorable for a change of their triplet electronic state. Therefore, this effect is the consequence of a nuclear *spin sorting process*.

Assuming chemically different radicals with different *g* values, the resulting nuclear spin polarization, and hence the type of the polarization pattern to be found in the CIDNP resonances of the reaction products can be predicted from the signs of four decisive variables according to a "sign rule" proposed by Kaptein (Eq. 1) [25, 26]:

$$\Gamma_n = \kappa \cdot \varepsilon \cdot g' \cdot a\,. \quad (1)$$

In this equation, κ describes the original spin state of the radical at birth (either singlet or triplet precursor) and ε stands for the type of reaction leading to the observed product (either cage recombination or cage escape); also important is the sign of g', the difference in g values of the radicals observed ($g' = g_1 - g_2$; where g_1 refers to the radical containing the nucleus being observed) and the sign of the hyperfine coupling constant a.

According to this rule, the phase (Γ_n) in which an individual resonance is observed in the CIDNP spectrum, i.e., whether it is observed in absorption (A) or in emission (E), is determined by the sign product of these four parameters (absorption, if the resulting sign is positive; emission, if the resulting sign is negative). A similar rule applies in situations where, in pairs consisting of two *chemically equivalent* radicals, either combination or cage escape products result. In these cases, the radicals may still contain magnetically active nuclei and hence be *magnetically inequivalent*. This also leads to the generation of CIDNP; however, the resulting polarization patterns are characterized by an equal number of emission and absorption lines in each multiplet (*multiplet effect*).

An even more complicated situation arises when free radicals are generated via an electron transfer. First of all, it is important to keep in mind that in this case either the donor or the acceptor molecules (or both) may be unsaturated and hence contain double bonds. If then, upon an electron transfer reaction, radical cations or anions result from such unsaturated systems, an electron back transfer can lead to the regeneration of the initial neutral molecules, i.e., to the starting materials. It will be virtually impossible to differentiate the regenerated molecules chemically from the originally present and yet unchanged starting materials, since both will be chemically identical. However, the molecules resulting from the electron back transfer may well be magnetically inequivalent relative to those initially present. Therefore, they can be distinguished via CIDNP using in situ NMR spectroscopy.

That this is true can easily be seen in the form of the observed nuclear spin polarization in the molecules of the starting materials. However, this electron back transfer might well exhibit the opposite spin selectivity because the law governing the formation of a single bond from a radical pair predicts that if an electron back transfer reconverts a radical cation into a neutral alkene, this may well be a viable process, even if it leads to the triplet state of the alkene. As a matter of fact, it is the energetics, in particular the change of the free enthalpy (ΔG), of the process that govern the relative rates of the two processes, i.e., the back transfer to the triplet state of the alkene or to its singlet state [27]. If it is energetically feasible to populate the triplet state of the alkene, this typically becomes the preferred route, resulting in a polarization pattern that has the opposite phase to that expected from the prediction of the aforementioned sign rule, which assumes an exclusive reactivity for the singlet reaction. In this case, the sign rule has to be expanded by one additional parameter to accommodate this new alternative, leading either to singlet or to triplet products, which may even be the starting materials themselves.

At this point, it is important to note that when characterizing free radicals via CIDNP, it is not the radicals themselves that are being detected. Rather, in this case, the information is gained indirectly from their final and hence stable diamagnetic reaction products, which are detected using NMR. Thereby, the intensity of the lines is higher, the more short-lived the radicals. Hence, CIDNP ideally complements ESR spectroscopy, which works better the more long-lived the radicals. The product-forming step is due to an encounter of a radical pair in a singlet state, which may occur either via combination or disproportionation of a geminate radical pair or a random radical pair. The latter case is outlined in Fig. 5.

$$^{\bullet}R^{\uparrow}\ ^{\bullet}R^{\downarrow} \longrightarrow R\text{-}R$$

Fig. 5 Product formation within radical pairs. Normally, the recombination product is formed during the encounter of a radical pair in a singlet state. This usually occurs via a combination or disproportionation step of a geminate radical pair or a random radical pair

In order to distinguish between these two alternatives, chemical or spectroscopic methods may be applied. Chemical methods make use of selective scavengers, which try to interfere with the product formation. The CIDNP mechanism, however, represents a spectroscopic method.

2.6.2 Molecular Dynamics (Definitions)

The nuclear polarization observed eventually in the products is a consequence of the spin selectivity that governs the possible reactions occurring in a radical pair. This makes it imperative to discriminate the different types of radical pairs, depending on the history of their formation:

- *Geminate radical pair*: A pair, whose partners are born (formed) together and share the same precursor molecule as a parent. A geminate pair may be born in the singlet state or in the triplet state. The latter applies if the cleavage occurs from a precursor molecule in its triplet state; a situation encountered frequently upon photochemical excitation of the precursor.

- *Random radical pair*: Random radical pairs are formed upon a random encounter of two free, i.e., uncorrelated, radicals. Random encounters yield predominantly pairs in their triplet state, which is three times as likely as the formation of a radical pair in its singlet state.
- *Free radicals*: Free radicals are uncorrelated radicals or radicals that have separated from a pair a sufficient distance such that non-geminate reaction with other partners is more likely than geminate reaction.
- *Solvent cage*: A solvent cage is made up of an ensemble of solvent molecules that surround and thereby enclose a molecule or a radical pair as the first shell, be it a geminate or a random radical pair.
- *Contact radical pair*: A contact radical pair comprises two radicals in a solvent cage without solvent molecules between them.
- *Solvent-separated radical pair*: A solvent-separated pair differs from a contact radical pair in so far as there are solvent molecules (one or more) between the two radicals. Solvent-separated pairs may occur as geminate or random radical pairs.

3 CIDNP Experiments

3.1 CIDNP During the Decomposition of Dibenzoylperoxide (DBPO)

3.1.1 ^{1}H & ^{13}C CIDNP During the Decomposition of DBPO

The thermal decomposition of dibenzoylperoxide (DBPO) represents a "classical" system, which was studied many years ago (Fig. 6). It was essentially the very first example where CIDNP was observed in a ^{1}H NMR spectrum [1, 2].

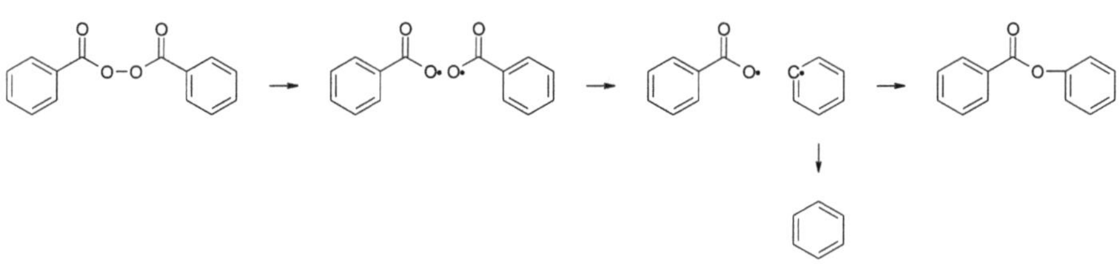

Fig. 6 Radical pair formation and generation of recombination (ester) and escape product (benzene) during the thermal decomposition of dibenzoylperoxide (DBPO). During the formation of the spin-correlated radical pair (step 2 to 3) one molecule of carbondioxide is cleaved from one of the benzoyloxy radicals yielding a radical pair whose radicals have different *g* values. Hence, the observation of the CIDNP *net effect* in the resulting CIDNP spectrum

If one compares both the ^{1}H-CIDNP and ^{13}C-CIDNP spectra recorded during the decomposition of DBPO in cyclohexanone at 90 °C using a 200 MHz ^{1}H NMR spectrometer with the "historical" spectrum recorded at 100 MHz it can be seen that there is little new information from the ^{1}H NMR spectrum when increasing the resonance frequency. Both the ^{1}H-CIDNP and ^{13}C-CIDNP spectra exhibit intense emission lines, which stem from benzene, the reaction product resulting from hydrogen abstraction from the solvent. In order to obtain detailed information about other products of this radical reaction, it proved much easier to study the thermal decomposition of the closely related system perfluorodibenzoylperoxide, which has been done using ^{19}F CIDNP.

3.1.2
^{19}F CIDNP During the Decomposition of ^{19}F-Substituted DBPOs

Many years ago, ^{19}F polarization in recombination and radical transfer products derived from pentafluorophenyl and monofluorophenyl radicals was observed [28]. On the basis of the spectral patterns, the signs of the hyperfine coupling constants for fluorine nuclei in the *ortho*, *meta*, and *para* position of these radicals could be determined.

When a solution of perfluorobenzoylperoxide in tetrachloromethane was photolyzed in the probe of an NMR spectrometer, seven enhanced multiplets were observed (Fig. 7). Three multiplets represented the phenoxyl segment of pentafluorophenyl pentafluorobenzoate: the *ortho* and *meta* nuclei appeared in enhanced absorption, and the *para* isomer in emission (E). A fourth, very weak multiplet was assigned to the *ortho* derivative of the benzoy-

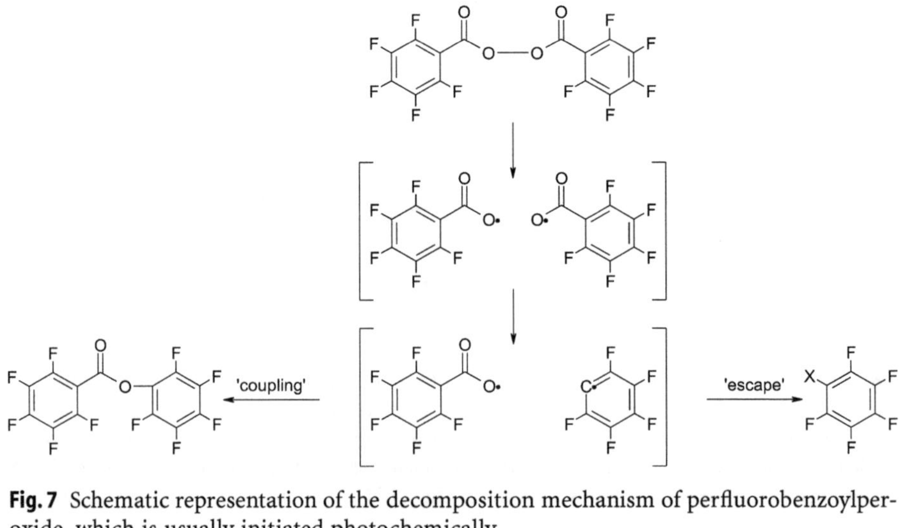

Fig. 7 Schematic representation of the decomposition mechanism of perfluorobenzoylperoxide, which is usually initiated photochemically

loxy fragment. The remaining three multiplets had the chemical shifts of the radical transfer product chloropentafluorobenzene with the opposite polarization to the former product. Decomposition of perfluorobenzoylperoxide in trichloromethane or in deuterated trichloromethane produced similar spectra and all signals showed only net polarization.

The investigators explained the observed polarization as being induced in the radical pair comprising the pentafluorobenzoyloxy radical and the pentafluorophenyl radical. The pentafluorobenzoyloxy radical should have a considerably larger *g* value than the pentafluorophenyl radical. If these apparently reasonable assumptions are correct, the signal directions observed for the coupling product and the radical transfer product indicate that the fluorine hyperfine coupling constants of the pentafluorophenyl radical are *positive* in the *ortho* and *meta* positions ($a_F^o, a_F^m > 0$), but negative for the *para* position ($a_F^p < 0$). From the intensity of the CIDNP spectra it appears that $a_F^o > a_F^m > \left|a_F^p\right|$. Because of possible relaxation effects the researchers did not attempt to determine the exact ratios of these coupling constants.

In addition, they studied the photodecomposition of ^{19}F-DBPO in tetrachloromethane. The polarized products observed in these reactions indicated that the fluorine hyperfine interactions of both *o*- and *p*-fluorophenyl radicals are positive and that a_F^o is larger than a_F^p.

In principle, the pattern of hyperfine interactions derived here for the pentafluorophenyl radical is very similar to the one found for the phenyl radical, i.e., $a_F^o > a_F^m > \left|a_F^p\right|$. Both experiment and calculation indicated that all three couplings are positive. CIDNP studies on *p*-chlorobenzoylperoxide and *p*-anisoyl peroxide confirmed that the hyperfine values for the *ortho* and the *meta* hydrogen nuclei are positive, while polarization has yet to be observed for the hydrogen nucleus in the *para* position.

A comparison of the pentafluorophenyl radical with ^{19}F-substituted π radicals, however, is illuminating. In π radicals, ^{19}F-substituents at the centers of high spin density can form π bonds to the aromatic system. As a result, the electron–fluorine interactions in both *ortho* and *para* positions are large and positive, whereas the hyperfine value for the *meta* ^{19}F nuclei is weaker and negative. Such a pattern is quite different from the one derived for the pentafluorophenyl radical, implying that this radical does not have appreciable π character. Accordingly, the theoretical understanding of fluorine hyperfine interactions appears to be less than adequate.

Nevertheless, the authors explain the large positive coupling of substituents in the *ortho* position by a direct interaction with the π orbital at C_1. For the *meta* and *para* positions the contributions of direct coupling will be smaller, and the hyperfine values of these nuclei will depend on a variety of factors including the polarization of the carbon-fluorine bond. Since it is conceivable that these contributions are sensitive to the nature of the substituents, the observed difference in the sign of a_F^p for the *p*-

fluorophenyl and the pentafluorophenyl radical does not appear internally inconsistent.

The researchers added a comment on the benzoyloxy radicals, which they and others have assumed as intermediates. They observed weak CIDNP signals for the *o*-fluorine nuclei of the benzoyloxy segments of the radical pair comprising the *o*-fluorophenyl radical and the *o*-fluorobenzoate radical. These signals indicate that the hyperfine coupling constants a_F^o of *o*-fluoro radicals and pentafluorobenzoyloxy radicals are positive. This limited information falls far short of elucidating the nature of ^{19}F-substituted benzoyloxy radicals satisfactorily.

3.2 CIDNP from Reversible Reactions

CIDNP can be a particularly valuable tool in the study of reversible reactions of free radicals where product and reagent are the same. We will provide details about a selection of reactions exhibiting this feature. To start, the decomposition of dibenzoylperoxide (vide supra) may serve as a characteristic example.

3.2.1 ^1H CIDNP During the Reversible Addition of Free Radicals to Aromatics

When aromatic molecules are present during the decomposition of ^{19}F-DBPO, CIDNP is usually observed in their NMR spectra. Figure 8 shows a typical ^1H CIDNP spectrum acquired during (upper part) and after (lower part) the thermal decomposition of a hexachlorobutadiene solution of ^{19}F-DBPO containing benzene. During decomposition, the benzene resonance line at $\delta = 7.4$ ppm is observed in emission, and other emission and enhanced absorption lines appear in other regions of the spectrum. The CIDNP of the substrate benzene indicates that some benzene molecules participate in a reaction during which they become part of a radical pair and consequently show a spin polarized pattern in the corresponding spectrum. We assume that the two radicals $R^\cdot$, obtained from the pyrolysis of ^{19}F-DBPO, add reversibly to aromatic molecules to afford ($C^\cdot$) of the cyclohexadienyl type. From the experimental results and the sign rule of CIDNP we derive the reaction sequence as outlined in Fig. 7.

Initially, the peroxide cleaves into a singlet radical pair $^1(R_1^\cdot R_2^\cdot)$. Apparently, one of the radicals adds to an aromatic molecule to yield a cyclohexadienyl radical ($C^\cdot$), which is still paired with one $R^\cdot$ in a singlet radical pair $^1(R^\cdot + C^\cdot)$. These singlet pairs can either form recombination or disproportionation products, or they can change their multiplicity to afford triplet radical pairs $^3(R^\cdot + C^\cdot)$. The triplet radical pairs will preferably separate and cleave into the reactants, i.e., two $R^\cdot$ and benzene.

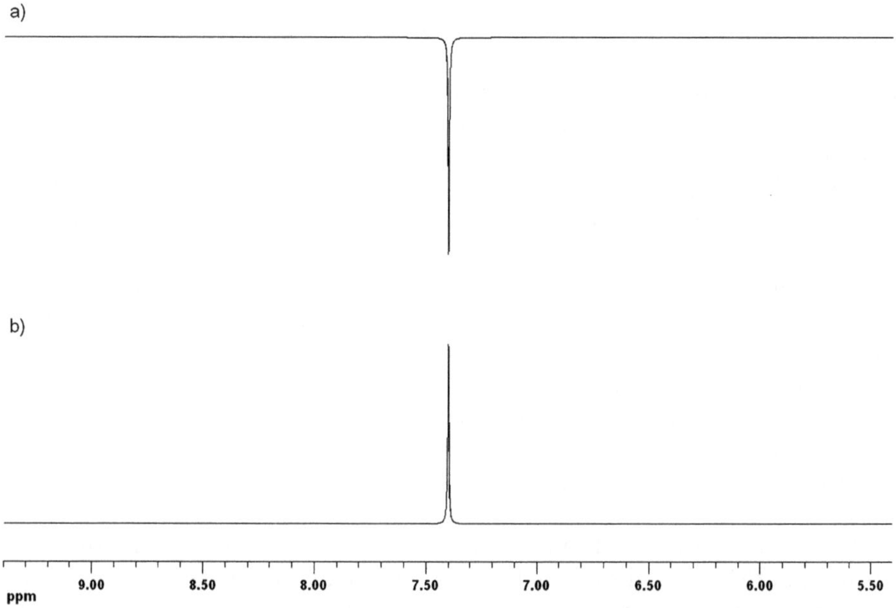

Fig. 8 ^{1}H-CIDNP NMR spectra acquired during (*a*) and after (*b*) the reversible addition of pentafluorobenzoyloxy radicals to benzene. The emissive line in *a* at $\delta = 7.4$ is the resonance line of benzene

Since the rates of multiplicity changes in radical pairs have been shown to be nuclear spin-dependent, CIDNP is expected in both the reactants and the products. We assume that $R^{\cdot}$ is a pentafluorobenzoyloxy radical. Therefore, the pairs $^1(R^{\cdot} + C^{\cdot})$ and $^3(R^{\cdot} + C^{\cdot})$ consist of radicals with different g values. Usually, radicals like $R^{\cdot}$ have higher g values ($g = 2.0045$) than cyclohexadienyl radicals ($g = 2.0025$). For $g(C^{\cdot}) < g(R^{\cdot})$ and for the magnetic fields usually applied, the rules of CIDNP predict "emission" for those product protons which experience negative hyperfine coupling constants (a_i), but "absorption" for the same protons in the reactants. Furthermore, an absorptive signal is expected for those product protons that experience positive hyperfine interactions in ($C^{\cdot}$), but the same protons should show an emissive signal in the reactants [29].

The $|a_i|$ of $C^{\cdot}$, generated by photolysis of R – R in the cavity of an ESR spectrometer, have been determined in the presence of benzene and CCl_4 or tetrachloroethane. Because $C^{\cdot}$ is an "odd alternant" carbon radical, we attribute the following signs: $a_1 = + 32.4 \pm 0.1G$, $a_2 = a_6 = - 8.8 \pm 0.1G$, $a_3 = a_5 = + 2.75 \pm 0.05G$, and $a_4 = - 12.95 \pm 0.1G$.

The parameters mentioned above explain the emission in the substrate benzene because of $|a_1 + a_3 + a_5| > |a_2 + a_4 + a_6|$. In the products, which give rise to the resonance lines at $\delta = 5.8$ ppm (absorption) and $\delta = 6.1$ ppm (emission), however, the positive a_i cause emission, and the negative a_i cause absorption.

1,3,5-Trichlorobenzene as the substrate shows emission, but only absorption is observed in the olefinic proton region at $\delta = 6.0$ ppm. This indicates that the free radical attack occurs at the unsubstituted carbons. Consequently, only positive a_i exist and this causes the observed emission in the substrate and absorption in the product molecules.

1,3,5-Tribromobenzene as the substrate, however, is observed in enhanced absorption. We conclude that here $g(C^{\cdot}) > g(R^{\cdot})$, so that the positive a_i now cause absorption in the substrate. This is consistent with observations of others that bromines attached to carbons with positive spin densities at C_2, C_4, and C_6 increase the g values of similar radicals [30].

Whereas three bromine substituents cause $g(C^{\cdot}) > g(R^{\cdot})$, two of them yield $g(C^{\cdot}) \sim g(R^{\cdot})$, because we observed the multiplet effect in *m*-dibromobenzene and similar substrates, in which two bromines were attached at positions C_2, C_4, or C_6.

The CIDNP patterns of the substrates allow us to assign the predominant position of the free radical attack in the substrates. For larger g values for $R^{\cdot}$ the protons at the "attacked" carbon atom will have the strongest emission, because a_1 is larger than all other a_i. Thus, in nitrobenzene the *meta* protons show emission, but the *para* and *ortho* protons show absorption. In anisole, however, the *ortho* and *para* protons show emission, but the *meta* protons show absorption. From similar results obtained with other monosubstituted benzenes we conclude that acceptor-substituted benzenes are attacked in the *meta* position, whereas donor-substituted benzenes are attacked in the *ortho* and *para* positions.

These substituent effects are those of electrophilic aromatic substitution, whereas in homolytic aromatic substitution all kinds of substituents favor attack at the *ortho* or *para* position. Therefore, the attacking radicals are not electrically neutral but are strong electrophiles. Consistently, in multisubstituted benzenes the protons *ortho* and *para* to the best donor always show emission and those *meta* show absorption.

These results have been obtained for a large number of aromatic substrates such as biphenyls, naphthalenes, and pyridines. Pyridine is attacked in the *meta* positions. In substituted pyridines the substituent effects are those of substituted nitrobenzenes. In general, donor-activated molecules are more easily attacked than acceptor-deactivated substrates. Alkyl side chains also show CIDNP, usually emission for the α-protons. Donors *ortho* or *para* to the alkyl groups cause side chain attack. In fluorinated substrates, ^{19}F CIDNP can be observed. However, from the proton CIDNP pattern one cannot always extrapolate the ^{19}F CIDNP spectra by means of the simple relationship $a_i^F = -|(Q_F/Q_H)|\, a_i^H$ [31, 32]. We found instead prevailing emission for fluorine, even in substrates like *p*-fluoroanisole, where we would have expected absorption. Similarly in α-fluorotoluene both the benzylic protons and the fluorine show emission, the three components of the fluorine triplet having intensity ratios differing from the absorption spectrum.

The ^{1}H CIDNP patterns of aromatic substrates are useful in the analysis of complex NMR spectra because adjacent protons become oppositely polarized (Fig. 9). The absolute signs of the NMR coupling constants can be determined from the multiplet effect, which prevails for all substrates at low magnetic fields.

Attempts to induce the same reactions by direct or sensitized photolysis of the peroxide have been unsuccessful so far. Therefore, we have not yet obtained direct evidence for the singlet character of the radical pair ($I^{\bullet} + R^{\bullet}$). However, the CIDNP results did not change when other free radicals were generated simultaneously, indicating that the radical pairs are not formed by secondary encounters of free radicals in the solution.

With 1,3,5-trichlorobenzene as the substrate, its resonance shows emission, and a lack of emission among the resonance lines of the products indicates radical attack at unsubstituted carbons, and thus only positive values for a_i. With 1,3,5-tribromobenzene, however, the substrate line shows enhanced absorption. Apparently, here the radical $C^{\bullet}$ has a higher g value than $R^{\bullet}$, which is expected if bromine replaces negatively coupled hydrogen nuclei. Interestingly, the two bromine nuclei of m-dibromobenzene increase the g value of its corresponding radical such that it matches that of $R^{\bullet}$, and the multiplet

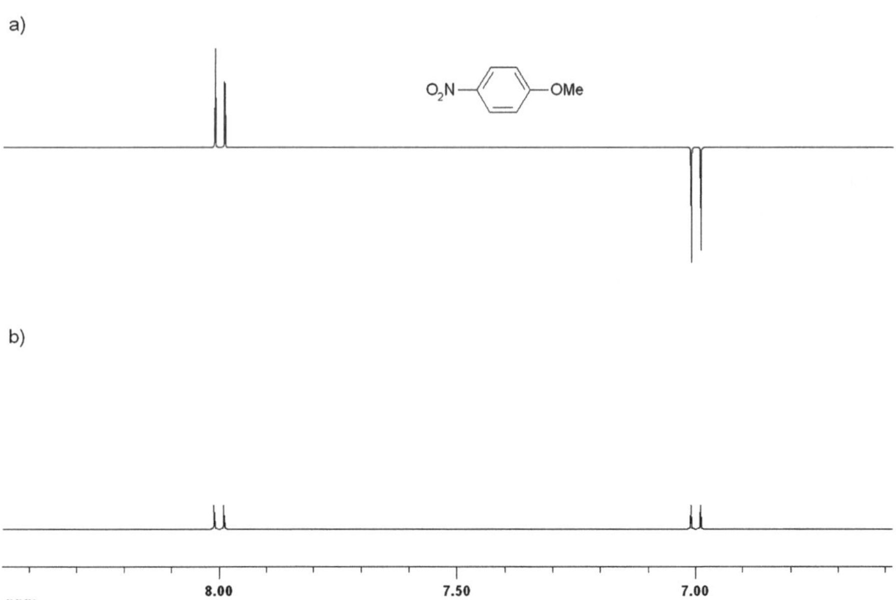

Fig. 9 ^{1}H-CIDNP NMR spectra acquired during (*a*) and after (*b*) the reversible addition of pentafluorobenzoyloxy radicals to 4-nitroanisole

effect (emission *and* absorption for the same proton) is observed. However, if bromine nuclei replace positively coupled hydrogen nuclei, they do not increase the *g* value of the radical. Correspondingly, *o*- and *p*-dibromobenzene show energy polarization (emission), and 1,2,4-tribromobenzene shows the multiplet effect. Because a_1 is the largest of all a_i, the strongest emission should result from the substrate protons attached to those carbon positions where the radical attack occurs. In multisubstituted benzenes the best donor substituents prevail in directing $R^{\bullet}$ into their *ortho* and *para* positions, even overruling other substituents and their contributions to resonance stabilization. In alkyl side chains, only the α-protons show CIDNP (emission). Donor substituents *ortho* or *para* to the alkyl groups favor side chain attack. Fluorinated benzenes show ^{19}F CIDNP, usually emission. Highly acceptor-deactivated molecules fail to show CIDNP in hexachlorobutadiene solution.

CIDNP is built up in σ complexes, thus giving no evidence for π complexes. It follows that CIDNP is useful in analyzing complex NMR spectra of aromatic molecules, because protons situated directly next to each other become oppositely polarized. Also, the absolute signs of the NMR coupling constants can be determined from the multiplet effect.

3.2.2 ^{13}C CIDNP During the Reversible Reaction of Pentafluorobenzoyloxy Radicals to Aromatics

CIDNP can be a particularly valuable tool in the study of reversible free radical reactions where product and reagent are the same. This type of reaction occurs during the decomposition of benzoyl peroxide in the presence of aromatic molecules. ^{1}H CIDNP studies of the decomposition of perfluorobenzoyl peroxide indicated that pentafluorobenzoyloxy radicals reversibly form σ complexes with a variety of aromatic substrates and that the site of attack is that expected for electrophilic aromatic substitution. We report the observation of CIDNP of ^{13}C in natural abundance during the photolysis of this compound in the presence of chlorobenzene. ^{13}C polarization observed during these reactions is of interest since the ^{1}H NMR spectra fail to give any information on the type of intermediate σ radical involved. Furthermore, it was found [33] that their decomposition in chlorobenzene gives phenyl pentafluorobenzoate in considerable yields, pointing to attack at the first position of the solvent. By contrast, the present CIDNP results indicate predominant addition to the carbon atoms in the second or the fourth position.

The ^{13}C FT NMR spectra were recorded on an NMR spectrometer modified such that the sample could be irradiated within the probe with the light of a high-pressure mercury arc [34]. When the light was admitted, the intensities of the carbons in the *ortho* and *para* positions increased, whereas those of the *meta* and quaternary carbons decreased. This polarization pattern can be accounted for by the reactions shown in Fig. 10.

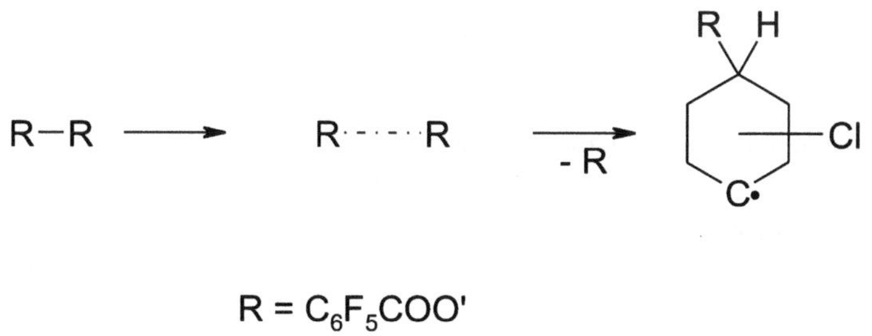

Fig. 10 Thermal decomposition of ^{19}F-DBPO and the subsequent addition of one radical to a chlorobenzene substrate molecule

The nature of the "recombination" step is not known; it may be coupling or disproportionation or both. Recombination is, however, a prerequisite for observing polarization in the escape product chlorobenzene. The spectrum suggests that addition occurs predominantly via attack at the *ortho* and *para* positions. This follows from an analysis of the polarization pattern in terms of the radical pair theory of CIDNP or, more conveniently, from the qualitative rules which summarize this theory.

Quantum chemical calculations at various levels of sophistication [35] predict alternating signs of the spin densities and, therefore, also alternating ^{13}C hyperfine couplings in the odd alternant cyclohexadienyl radical in accordance with the observed spectra (Fig. 11). In an odd alternant radical, the contributions from neighboring atoms to the hyperfine coupling constant of a carbon nucleus have signs opposite to that of the main carbon atom, thus reinforcing the alternating effect in the ^{13}C hyperfine couplings. The *meta* adducts and related species would yield the opposite polarization. Thus, the present results give no indication of attack at the first position of chlorobenzene to yield the σ complex which, as has been proposed by Oldham et al. (vide supra), would explain the formation of phenyl pentafluorobenzoate. Either this product is generated via a different route or the radical is indeed formed but not reversibly.

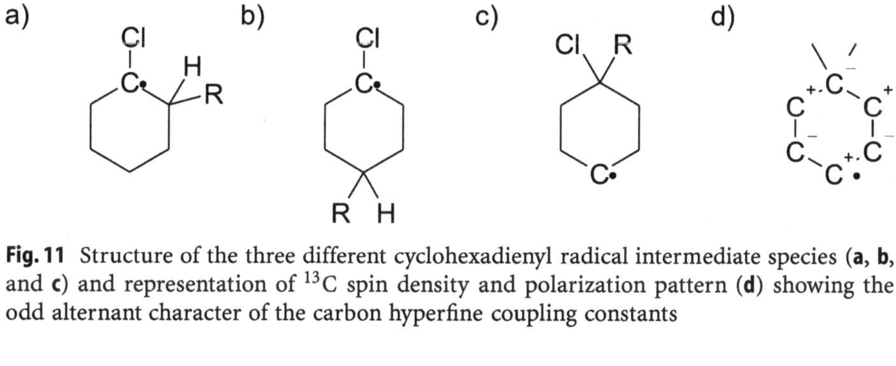

Fig. 11 Structure of the three different cyclohexadienyl radical intermediate species (**a**, **b**, and **c**) and representation of ^{13}C spin density and polarization pattern (**d**) showing the odd alternant character of the carbon hyperfine coupling constants

Table 1 Reversible radical reactions suitable for probing various chemical systems and substrates

Compound/functionality	Method/system	Remarks
Aromatic components	(a) Reversible addition of oxyradicals (b) Photooxidation (c) Photoreduction	Not suitable for – OH, – SH, – NH_2, – CH_3, Alkyl, – F, – Cl, – Br, – I
Alcohols	Enols from carbonyls	Also via photoreduction; mostly suitable for aliphatics
Olefins	Electron transfer & isomerization	Cleavage of halides
Amines	Reversible electron transfer	Not suitable for aromatic amines
Aldehydes	Reversible cage return	Stereochemistry altered
Ketones	Reversible hydrogen abstraction	Works both for protons and carbon nuclei
Phenols	Reversible hydrogen abstraction	Works both for protons and carbon nuclei

Observation of the light-induced intensity differences of the relatively strong solvent lines was greatly facilitated by making use of a technique for the suppression of the NOE [36]. The ^{1}H decoupler was gated such that it was "on" only during acquisition of the transients (0.8 s) and "off" during a waiting time (15 s) between the $\pi/2$ pulses. This procedure results in elimination of the NOE enhancement while the ^{1}H decoupling effect is retained. The effects were qualitatively the same when the decoupler was "on" continuously. Hence, the observed phenomena are not caused by an indirect ^{1}H – ^{13}C relaxation effect due to polarized protons, because the decoupling power is large enough to "short circuit" the relaxation transitions.

A variety of other radical reactions, most of them reversible, have been identified that are suitable for elucidating the structure and functionality of enzymes via CIDNP (Table 1). The most prominent system is based on the reversible hydrogen abstraction of photoexcited dyes or ketones from phenolic side groups in proteins. Likewise, this system qualifies for investigating simple ketones or phenols both using ^{1}H or ^{13}C CIDNP, and it provides access to the signs of coupling constants that can otherwise be determined from regular NMR spectra only with considerable difficulties [37–39].

3.3 CIDNP of Biomacromolecules

In 1978 it was demonstrated that CIDNP, generated in photochemical reactions between an excited dye and certain amino acid side chains present on the surface of a protein, could be used to monitor the solvent accessibility or exposure of these residues. This "photo-CIDNP" technique has proved a powerful probe of protein dynamics and of the wide variety of factors that modify the accessibility of the polarizable amino acid residues. Applications include studies of the interaction of proteins, and investigation of conformational changes and denaturation. In the first authoritative reviews of the biological applications of photo-CIDNP published in the early 1980s it was mentioned that the method is still very much in the process of being developed and that these surveys should best be regarded as a progress report. In the time since then a lot of progress on both the methodology and the number of possible applications has actually been achieved.

Typically, the "biological" photo-CIDNP method is applied for structure determination of proteins and small peptides. For this purpose, one takes advantage of the fact that the solvent accessibility of the amino acid side-chains of tyrosine (Tyr), histidine (His), and tryptophan (Trp) characteristically differ for an induced electron transfer reaction from acceptor dyes (Fig. 12). This makes it possible to assign the NMR resonances of different side-chain nuclei of these types. Upon the application of laser light, the accessible aromatic amino acid side-chains of a protein exhibit nuclear polarization if, for example, polyaromatic dyes like 3-*N*-carboxymethyllumiflavin (Flavin 1) (Fig. 13) are used. These small molecules act as photosensitizers that electrochemically catalyze a cyclic radical reaction scheme, which gives rise to the CIDNP effect in amino acids and proteins (Fig. 14).

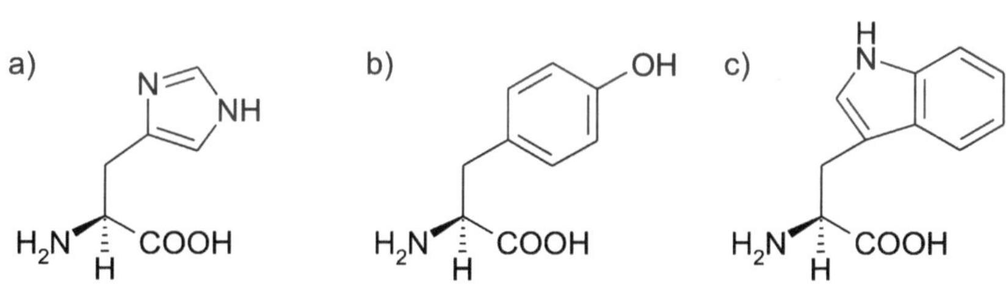

Fig. 12 Schematic representation of the three aromatic L-amino acid side chains histidine (**a**), tyrosine (**b**), and tryptophan (**c**) that are potentially CIDNP-active

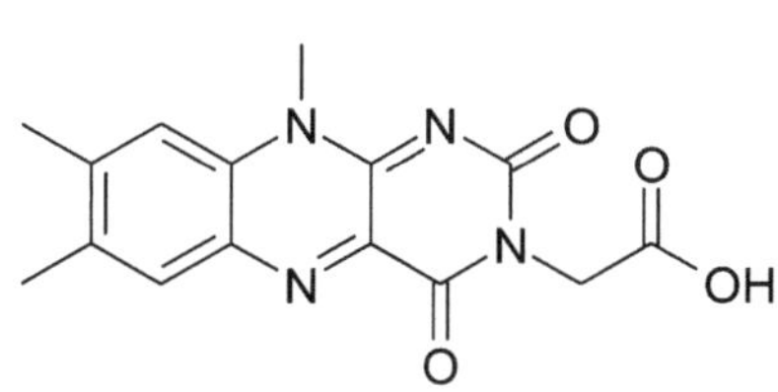

Fig. 13 3-*N*-Carboxymethyllumiflavin, a typical dye of the flavin family, used as a photosensitizer for the determination of the structure of amino acids and proteins via photo-CIDNP

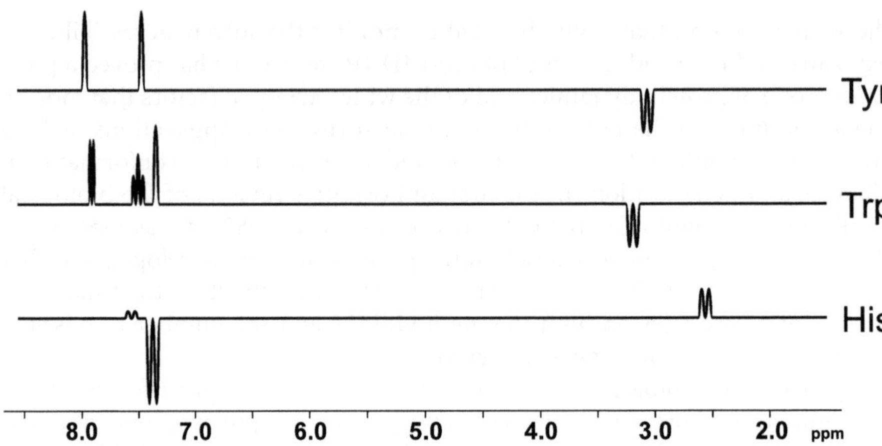

Fig. 14 Schematic representation of the ^{1}H CIDNP spectra of the amino acids tyrosine (Tyr), tryptophan (Trp), and histidine (His)

In spite of the success of photo-CIDNP of biomacromolecules and, in particular, of proteins using acceptor-type dyes of the flavin family [40], there is still a lack of and hence a need for alternative approaches to highlight certain sections or accessible side-chains for CIDNP-based structure determination of proteins. We want to suggest alternative approaches to this particular goal that take advantage either of the nuclear Overhauser effect (NOE) or use reversible photochemical reactions such as reversible addition of free radicals, hydrogen abstraction, or electron transfer.

3.3.1 Transfer of CIDNP via Proton Exchange and Nuclear Overhauser Effect (CINOE)

It is typically assumed that if a molecule displays CIDNP, the polarized nuclei must have experienced a hyperfine coupling in an intermediate paramagnetic state themselves. However, CIDNP can also become transferred in an *intermolecular* fashion, through a chemical exchange of spin-polarized protons. In this case, the NOE causes a secondary nuclear spin polarization in all those nuclei that are coupled by either scalar (negative NOE) or dipolar (positive NOE) interactions to the exchanging protons. Consequently, seemingly inert molecules may display CIDNP deceptively; whereas this phenomenon is in fact a "chemically induced" (or "pumped") NOE (termed CINOE hereafter). The latter may complicate the analysis of CIDNP spectra, but it may also be exploited to advantage for the following applications: (i) selective signal enhancement of protons and of other nuclei (^{13}C, ^{15}N); (ii) assignment assistance for specific resonances, and (iii) study of proton exchange and the dynamics thereof.

The concept of CINOE follows readily from the polarization transfer experiments together with the known negative or positive NOE associated with chemical exchange of spin-polarized protons. Whereas an NOE is typically pumped electronically, the CINOE approach offers certain advantages: (i) the initial polarization can be orders of magnitude larger than that generated electronically; (ii) the sign of the transferred polarization can conveniently be chosen to be either positive or negative by changing one of the parameters connected by the sign rules of CIDNP [11] or subsequent additions thereto [9]; (iii) broad resonances can also be pumped.

The latter allows pumping of the broad resonances of protons attached to ^{14}N (or any other quadrupolar nucleus), since only the T_2 time and not the T_1 time of these protons is short. Electronically, such broad resonances are very difficult if not impossible to pump. This can be seen looking at the results for formamide and for the uniformly ^{15}N-labeled formamide species. In addition to the NMR spectra [9], the CINOE spectra were obtained using equimolar mixtures of biphenyl-d_{10} and triethylamine (TEA) in CD_3CN as a solvent.

A prerequisite for CINOE studies, namely an efficient pump procedure, was found in photoinduced electron-transfer reactions between tertiary aliphatic amines as donors and a variety of acceptors (A) listed in Table 2. The α-protons, i.e., the protons attached to the α-carbons of the tertiary amine-derived radical cations, are acidic and exchange readily with the marked protons (– H). Many reactions that generate aminium cations and their associated CIDNP have been outlined previously.

These photoinduced electron-transfer reactions require polar solvents such as acetonitrile or dimethyl sulfoxide. The acceptors may be (i) aliphatic or aromatic nitriles, (ii) aromatic hydrocarbons, (iii) ketones or quinones. Table 2 lists the sign (phase) of CIDNP observed in the α-protons of the amines (– H″) and of the exchanging protons (– H), which is always opposite.

Table 2 Observation of CINOE observed using different substrates, acceptors, and donors

Acceptor	Donor	Polarized in …		
		– H	– H′	– H″
Biphenyl	Triethylamine	E	E	A
Naphthalene	Triethylamine	E	E	A
Naphthalene	PDM-amine	E	E	A
Naphthalene	PDM(d_8)amine		none	
Naphthalene	*N*-Alkylpyrrole	A	A	E
CD_3CN	Triethylamine	A	A	E
Benzonitrile	Triethylamine	A	A	E
Benzonphenone	Triethylamine	A	A	E
Anthraquinone	Triethylamine	A	A	E

E emission, A absorption

The chemistry of the CIDNP pumping as well as of the associated NMR and CIDNP lines should be compatible with the substrate. Some overlap of resonances is acceptable.

The pump should also be highly reversible to assure a long duration of the CINOE. Nitriles and hydrocarbons were found to yield a higher reversibility than quinones or ketones. The irreversibility could be shown to be due to photoreduction of A to AH_2 and simultaneous oxidation (and possibly hydrolysis) of the amine (Fig. 15). In support of the above scheme, the CIDNP resonances of the reduction and oxidation product are observed. Furthermore, the polarization in the positions – H, – H′, and H″ disappears upon deuteration of the α-positions in the amines. In the presence of suitable bases, even the solvent CD_2HCN displays CINOE because of the exchange equilibrium between the methoxide and the deuterated acetonitrile. Water can become strongly polarized, especially when occurring in small concentrations in aprotic solvents. This fact may be used to pump aqueous systems via a membrane together with a flow system. The molecules investigated here are models for those of biochemical interest such as proteins, sugars, or nucleic acids. There, the CINOE might well be quite selective. Especially attractive seems the possibility of pumping ^{15}N, where electronic pumps are inefficient because of the negative magnetic moment of ^{15}N and the associated negative enhancement, which causes an initial signal loss. CINOE could provide a superior alternative for positive ^{15}N signal enhancement. The above ^{1}H results seem very encouraging.

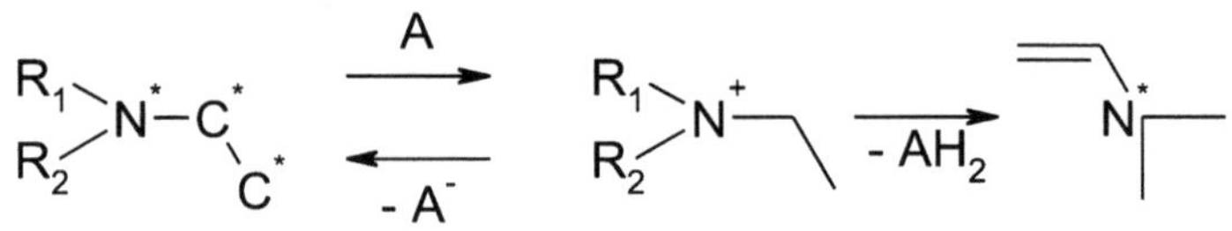

Fig. 15 Generation of spin-polarized exchangeable protons for CINOE

The systems, as outlined here and applied to small molecules, qualify for highlighting or probing accessible functional groups in proteins or enzymes in an approach equivalent to that used originally. The CINOE phenomenon itself has over the years been reported on by a number of independently working researchers [41–46].

4 Concluding Remarks

Photo-CIDNP NMR spectroscopy represents an active field of research both in chemistry and, in particular, in biochemistry. In this review, we have outlined the historical development of the technique and its importance for sensitivity enhancement during the observation and investigation of "free radical" reactions in solution using NMR spectroscopy. We have shown how to predict the phase of the chemical enhancement using a simple rule that correlates the signs of the four relevant magnetic parameters of a radical pair with the absolute phase of the spectroscopic amplification induced by CIDNP. Furthermore, we have discussed the dominating mechanism believed to be responsible for the CIDNP phenomenon itself and we have given examples of how CIDNP can be used to obtain important information about the mechanisms of several chemical reactions, including the thermal decomposition of aromatic and aliphatic peroxides and the reversible addition of free radicals to simple aromatic compounds. The review was concluded by a description of how "biological" CIDNP experiments are generally conducted and what information about the structure of proteins and other biomacromolecular complexes can be gained from them. We strongly believe that in this sense the photo-CIDNP in situ NMR technique will find an even greater variety of important applications in the future. These will undoubtedly further promote the general applicability of the method.

Acknowledgements L.T.K. thanks the German National Academic Foundation (Studienstiftung des deutschen Volkes) for constant support and generous funding throughout his studies. Further financial support from the Deutsche Forschungsgemeinschaft (DFG) and the Theodor-Laymann-Stiftung (L.T.K.) is also gratefully acknowledged.

References

1. Bargon J, Fischer H, Johnsen U (1967) Z Naturforschg A 20:1551
2. Bargon J, Fischer H (1967) Z Naturforschg A 20:1556
3. Ward HR, Lawler RG (1967) J Am Chem Soc 89:5518
4. Kaptein R, Dijkstra K, Nicolay K (1978) Nature 274:293
5. Hore PJ, Broadhurst RW (1993) Prog Nucl Magn Reson Spectrosc 25:345
6. Mok KH, Hore PJ (2004) Methods 34:75
7. Kuprov I, Hore PJ (2004) J Magn Reson 171:171
8. Fessenden RW, Schuler RH (1963) J Chem Phys 39:2147
9. Eichwald C, Walleczek J (1997) J Chem Phys 107:4943
10. Ritz T, Adem S, Schulten K (2000) Biophys J 78:707
11. Ritz T, Dommer DH, Phillips JB (2002) Neuron 34:503
12. Ritz T, Thalau P, Phillips JB, Wiltschko R, Wiltschko W (2004) Nature 429:177
13. Closs GL (1969) J Am Chem Soc 91:4552
14. Closs GL, Trifunac AD (1969) J Am Chem Soc 91:4554
15. Closs GL, Trifunac AD (1970) J Am Chem Soc 92:2183
16. Closs GL, Trifunac AD (1970) J Am Chem Soc 92:2186
17. Closs GL, Doubleday CE, Paulson DR (1970) J Am Chem Soc 92:2185
18. Kaptein R, Oosterhoff CJ (1969) Chem Phys Lett 4:195
19. Kaptein R, Oosterhoff CJ (1969) Chem Phys Lett 4:214
20. Adrian FJ (1970) J Chem Phys 53:3374
21. Adrian FJ (1971) J Chem Phys 54:3912

22. Noyes RM (1954) J Chem Phys 22:1349
23. Closs GL, Czeropski MS (1977) J Am Chem Soc 99:6127
24. Bargon J (1977) J Am Chem Soc 99:8350
25. Kaptein R (1971) Chem Comm, p 732
26. Kaptein R (1972) J Am Chem Soc 94:6251
27. Bargon J (2006) Photochem Photobiol Sci 5:970
28. Roth HD, Kaplan ML (1973) J Am Chem Soc 95:262
29. Bargon J (1971) J Am Chem Soc 93:4630
30. Sinclair J, Kivelson D (1968) J Am Chem Soc 90:5074
31. McConnell HM (1956) J Chem Phys 24:764
32. Maki AH, Geske DH (1961) J Am Chem Soc 83:1852
33. Oldham PH, Williams GH (1970) Chem Comm, p 1260
34. Kaptein R, Freeman R, Hill HDW, Bargon J (1973) Chem Comm, p 953
35. Pople JA, Beveridge DL, Dobosh PA (1968) J Am Chem Soc 90:4201
36. Freeman R, Hill HDW, Kaptein R (1972) J Magn Reson 7:82
37. Maurer HM, Gardini GP, Bargon J (1979) Chem Comm, p 272
38. Maurer HM, Bargon J (1979) J Am Chem Soc 101:6865
39. Maurer HM, Bargon J (1980) Org Mag Res 13:430
40. Mok KH, Nagashima T, Day IJ, Hore PJ, Dobson CM (2005) Proc Natl Acad Sci USA 102:8899
41. Closs GL, Czeropski MS (1977) Chem Phys Lett 45:115
42. Closs GL, Czeropski MS (1978) Chem Phys Lett 53:321
43. Kaptein R, Dijkstra K, Mueller F, Van Schagen CG, Visser AJWG (1978) J Magn Reson 31:171
44. Roth HD (1973) Mol Photochem 5:91
45. Roth HD, Manion ML (1975) J Am Chem Soc 97:6886
46. Kuprov I, Hore PJ (2004) J Magn Reson 168:1

Author Index Volumes 251–276

Author Index Vols. 26–50 see Vol. 50
Author Index Vols. 51–100 see Vol. 100
Author Index Vols. 101–150 see Vol. 150
Author Index Vols. 151–200 see Vol. 200
Author Index Vols. 201–250 see Vol. 250

The volume numbers are printed in italics

Subject Index

MIX
Papier aus verantwortungsvollen Quellen
Paper from responsible sources
FSC® C105338

If you have any concerns about our products,
you can contact us on
ProductSafety@springernature.com

In case Publisher is established outside the EU,
the EU authorized representative is:
Springer Nature Customer Service Center GmbH
Europaplatz 3, 69115 Heidelberg, Germany

Printed by Libri Plureos GmbH
in Hamburg, Germany